Biological Scanning Electron Microscopy

Biological Scanning Electron Microscopy

Barbra L. Gabriel

VAN NOSTRAND REINHOLD COMPANY
NEW YORK CINCINNATI TORONTO LONDON MELBOURNE

Library of Congress Catalog Card Number: 82-2695
ISBN: 0-442-22922-4

Manufactured in the United States of America

Published by Van Nostrand Reinhold Company Inc.
135 West 50th Street, New York, N.Y. 10020

Van Nostrand Reinhold Publishing
1410 Birchmount Road
Scarborough, Ontario MIP 2E7, Canada

Van Nostrand Reinhold Australia Pty. Ltd.
17 Queen Street
Mitcham, Victoria 3132, Australia

Van Nostrand Reinhold Company Limited
Molly Millars Lane
Wokingham, Berkshire, England

15 14 13 12 11 10 9 8 7 6 5 4 3 2 1

Library of Congress Cataloging in Publication Data

Gabriel, Barbra.
Biological scanning electron microscopy.

Includes index.
1. Scanning electron microscope. I. Title.
QH212.S3G3 578′.45 82-2695
ISBN 0-442-22922-4 AACR2

To AMG and ARG

Preface

The past eighteen years has seen the Scanning Electron Microscope evolve from an esoteric research instrument into a familiar, but still exceptional, implement. Parallel with the theoretical and technological developments of the microscope were those techniques pertaining to the preparation of biological specimens in such a manner that they duplicated what was observed in life. By drawing upon the extensive knowledge accumulated by light and transmission electron microscopists, scanning microscopists adopted, modified, and developed new methods applicable to this new technology. It is hoped that these following pages reveal the magnitude of effort that innumerable researchers have expended to increase our knowledge.

These preparation methods are discussed in both theoretical and practical terms, combining an historical atmosphere and culminating in state-of-the-art techniques. Although these are basic methods, their fundamental nature does not diminish their importance. Undoubtedly, future developments will revise, and possibly replace, these techniques. The author has attempted to give full credit to the researchers who provided the entire framework to this text; for those who were inadvertently not cited, I sincerely apologize for the omission.

Barbra Gabriel

Acknowledgments

My gratitude extends to my colleagues who shared their knowledge, offered encouragement, and judicially critiqued my work. Special thanks to my reviewers, Dr. Fredric Giere and Dr. Bruce Murray, and to the editors of Van Nostrand Reinhold for their guidance.

Contents

Preface vii

1. SCANNING ELECTRON MICROSCOPE **1**

Illuminating System 3
Information System 9
Detection System 12
Display System 13
Vacuum System 14
Resolution 19
Practical Interference Problems 29
Beam-Specimen Interactions 30

2. SEM PHOTOGRAPHY **33**

Introduction 33
Exposure 34
Types of Film 38
Printing 40

3. INTRODUCTION TO BIOLOGICAL SAMPLE PREPARATION **43**

Specimen Drying 43
Specimen Mounting 46
Coating Techniques 47

4. CHEMICAL FIXATION **50**

Chemical Fixatives 51
Organic Dehydration 79

5. CRITICAL POINT DRYING 96

Theory 96
CPD Apparatus 101
Method 104

6. FREEZE-DRYING 108

Comparison of FD and CPD 112

7. SPECIMEN DRYING FROM VOLATILE REAGENTS 116

8. HANDLING FREE-LIVING CELLS 123

Macroscopic Organisms 123
Microscopic Cells and Organisms 124

9. CONDUCTIVE THIN FILMS 131

Introduction 131
Evaporated Thin Films 131
Method of Evaporation 132
Sputtered Films 139
Comparison of Evaporation and Sputtering 143

10. UNCOATED SPECIMENS 148

Fresh or Frozen Specimens 149
Metallic Impregnation 150
Summary 160

Appendix 165

Index 173

Biological Scanning Electron Microscopy

1. Scanning Electron Microscope

The scanning electron microscope (SEM) was commercially introduced in the mid-1960s, although the basic construction was proposed as early as 1935. Current instrumentation routinely guarantees a 70 Å resolution, which is a significant improvement over the 250 Å resolution of 10 years ago. What distinguishes the SEM from any other microscope is that it is capable of an unusually great depth of focus. Consequently, SEM images impress the observer with a seemingly three-dimensional structure.

The major purpose of the SEM is for morphological surface analysis. This implies that information is "reflected" from the sample surface, much as one views images by reflectance optical microscopy. The major difference between these two systems is that the optical microscope reflects photons, whereas the SEM analyzes electrons released from the specimen.

Because electrons are relatively energetic particles, they may also induce other information signals within a sample, including cathodoluminescence and x-rays. Consequently, other developments which are direct outgrowths of the SEM include energy-dispersive x-ray spectroscopy and detection of cathodoluminescence by photomultipliers.

The SEM basically consists of five systems (Figure 1-1):

1. Illuminating/imaging system, consisting of the electron gun and magnetic lenses.
2. Information system, or sample, which releases a variety of information signals.

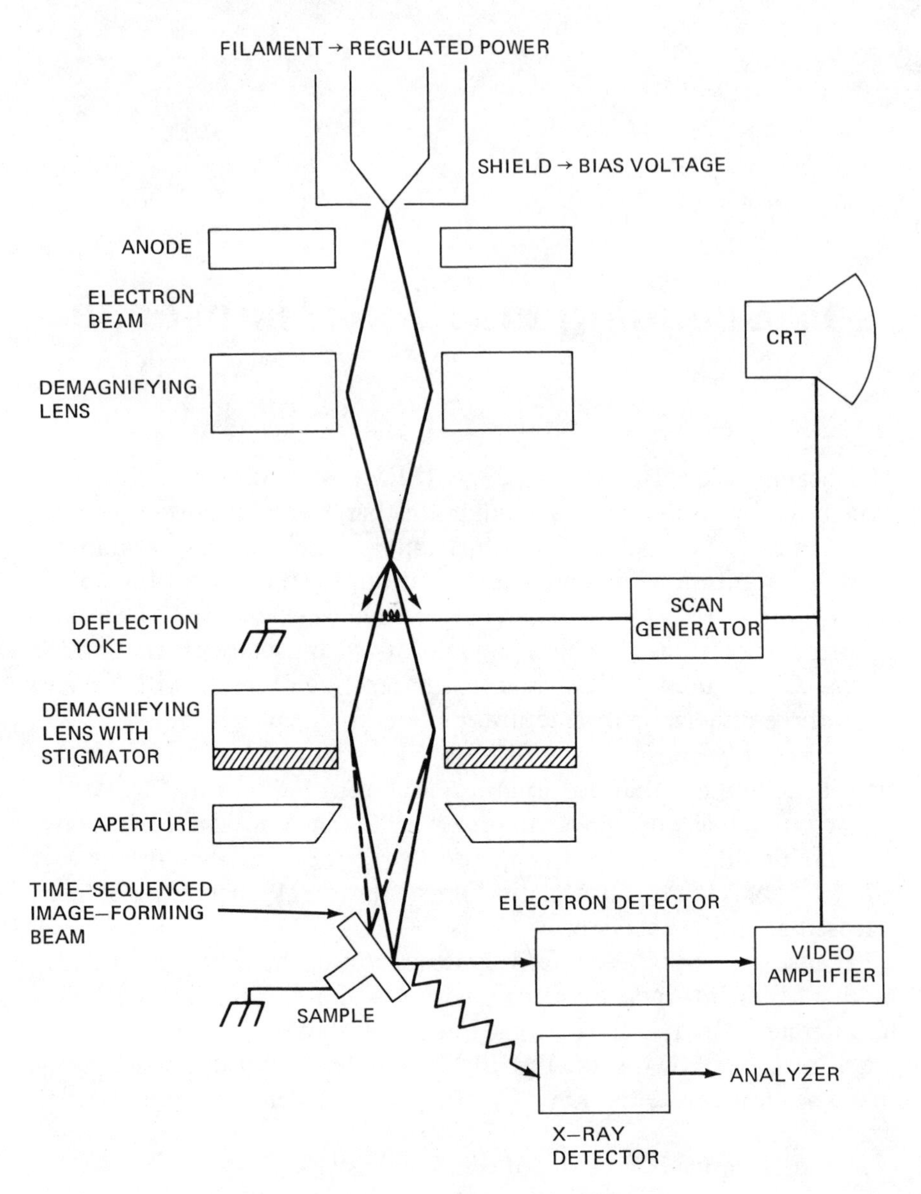

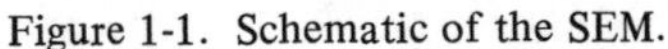
Figure 1-1. Schematic of the SEM.

3. Detectors, to recognize and analyze the emitted signals.
4. Display system, to faithfully reproduce the detected signals.
5. Vacuum system, to remove gases from the column.

ILLUMINATING SYSTEM

The illuminating system serves to produce the electron beam and focus it onto the sample (Figure 1-2). It consists of an electron gun which generates electrons, and a demagnifying (condenser) lens assembly to direct the beam onto the sample. Together, these components have the properties of, and act like, an electrostatic lens.

The electron gun has three components: (1) a filament, or electron

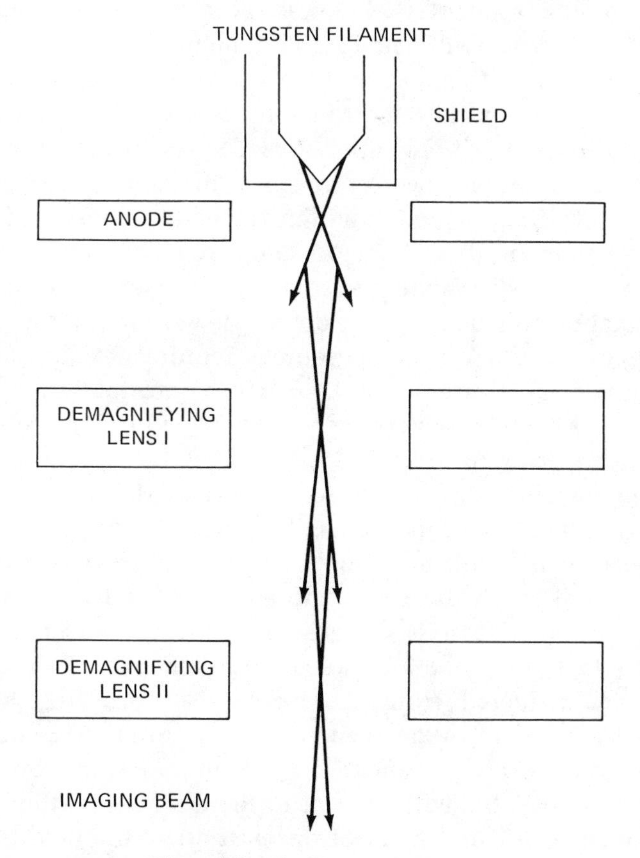

Figure 1-2. The illuminating system.

source, held at a negative potential, (2) a shield at a slightly positive potential, and (3) an anode held at a large positive potential with respect to the filament. Applying a sufficiently high current through the filament produces electrons, which are then attracted down the microscope column by the potential difference of the anode.

The prerequisities for the filament are that it be small, symmetrical, and capable of producing an intense beam of electrons. Most conventional microscopes meet these requirements by using a V-shaped tungsten wire, about 0.2 mm in diameter, which can be electrically heated to incandescence. The filament is held at approximately 10,000 volts below ground potential, and is thus the equivalent of a cathode. The effective electron source, or operating temperature, is referred to as saturation. Heating beyond this point will severely decrease the life span of it; likewise, admitting air without a few minutes of cool-down will damage the filament.

A heated tungsten filament is the conventional electron source for most SEMs; other special electron sources are the lanthanum hexaboride (LaB_6) and the field-emission (FE) guns. Both of these latter sources produce bright beams of small diameter and have much longer lifetimes than heated tungsten filaments. LaB_6 and FE guns require a vacuum of, 1×10^{-6} torr and 1×10^{-10} torr, respectively, at the gun level. The vacuum system of a conventional SEM is not usually capable of reaching this degree of vacuum at the gun level; these nonconventional electron sources require an additional high-vacuum pump at the gun level. Further information concerning LaB_6 guns is available in Verhoven and Gibson (1976, 1977) and FE guns are discussed by Welter (1975).

Enclosing the filament is the second part of the electron gun, the shield (synonyms: grid cap, Wehneldt cylinder). It is a cylinder, having an aperture 1–3 mm in diameter, that must be centered over the filament tip. The distance separating the tip and aperture greatly influences beam brightness, because a relatively small potential difference, or bias, is applied between the filament and shield. Thus, electrons are attracted from the filament toward the shield, and accelerated from here by an even-greater potential difference toward the anode. Therefore, the anode is at ground potential, but at a very large negative potential with respect to the filament (cathode). After leaving the electron gun, the electrons pass down the field-free region of the column at constant velocity.

The electron gun is also responsible for the accelerating voltage (synonym: accelerating potential) of the SEM. This is the energy with which electrons travel down the column; it is expressed in kiloelectron volts, or keV. Most SEMs are equipped with varying accelerating voltages of 5, 15, and 30 keV; other instruments may also have a 50-keV option. As will become clear when beam/specimen interactions are discussed, the higher the accelerating voltage, the greater the depth of penetration by the beam. In general, most biological samples prepared by conventional means–*i.e.*, having a conductive coating–are examined at 15 keV. Nonconductive samples are examined at 1–5 keV to avoid a common problem referred to as "charging," which is accumulation of electrons by the sample. When charging occurs, the resolution decreases because the specimen deflects the electron beam (like charges repel one another). Although there are no radical improvements in resolution by increasing accelerating voltage, it has been proven theoretically that wavelength, which controls resolution, is itself controlled by accelerating voltage (*i.e.*, higher voltage $\longrightarrow$ shorter wavelength $\longrightarrow$ higher resolution). This topic will be discussed later.

The second component of the illuminating system is the demagnifying-lens assembly. It serves the dual function of demagnifying the electron beam (Figure 1-2) and focusing it onto the sample. The beam diameter leaving the electron gun is 25,000–50,000 Å, but is reduced to about 100 Å after being acted upon by the lens assembly. The final lens is also equipped with a movable aperture which intercepts excess electrons–thereby preventing background scattering and thus reducing noise in the signal. Consequently, the effective electron gun current of about 10^{-4} amp (roughly the equivalent of 10^{15} electrons/sec) is reduced to 10^{-12}–10^{-10} amp (or 6×10^{6} electrons/sec) at the sample.

Changing the focal length of the demagnifying lens controls brightness. The SEM, in addition to the brightness and contrast controls which determine image appearance on the cathode-ray tube (CRT), has a control for spot size (synonym: beam diameter). This regulates the diameter of the beam as it strikes the sample and sets the maximum resolution for a given sample. Assuming a "perfect" sample–for example, if the beam diameter is 250 Å–the resolution cannot be better than 250 Å. Simple observation of a specimen may be accomplished with a large spot size, but high-resolution photography requires that the spot size be reduced. When reduced, the image observed on TV scan will be grainy, but at visual scan on the recording CRT the image is of better resolution.

Most tungsten hairpin filaments have a lifetime of at least 40 hr, when operating correctly: Correct operation requires that the user turns the filament on at a vacuum of at least 10^{-4} torr; does not exceed saturation; and waits a few minutes after turning the filament off and bringing the column to atmospheric pressure. Otherwise, gas will erode the hot filament.

Specifications for changing a filament are included with the manufacturer's operating manual for a given microscope, but a few general rules apply:

1. Always wear lint-free cotton or nylon gloves when handling the gun assembly. Also necessary are cotton-tipped applicators, Kimwipes, oil-free acetone (*e.g.*, EM grade), compressed Freon, and metal polish.
2. The inside of the shield usually shows a bluish discoloration from the evaporation of tungsten, or a black discoloration from contamination. This is cleaned as follows.
   a. Using a commercially available metal polish and a Kimwipe, buff the shield until it is free from discoloration. Also clean the aperture, and any contamination from the outside of the shield, with a cotton applicator.
   b. Remove the excess polish with acetone, paying particular attention to the aperture and heat vents.
   c. If available, place the shield in a small beaker of acetone and then into an ultrasonic bath.
   d. Handle the shield with gloves only.
3. Observe the burned-out filament with a binocular light microscope.
   a. A rough, eroded surface is typical, even after a reasonable lifetime. If the filament burned out after less than 15 hr of use, check for vacuum leaks.
   b. If beads are observed at the broken ends, tungsten is melting—meaning that a high-voltage problem exists.
4. The filament electrodes, shield, and all exposed parts of the gun should be cleaned with acetone followed by compressed gas before reassembly; and all debris cleaned out after reassembly.

5. Never leave a gun assembly open to the atmosphere, as severe contamination is inevitable. Do not stop anywhere in this procedure; continue through as one operation.

Traditionally, the final lens of the illuminating system is usually referred to as the objective lens. Actually, however, it functions exactly like a demagnifying lens, in that it demagnifies the electron beam. It has a number of features which the condenser does not—*e.g.*, it focuses the electron beam, which is accomplished by varying the current that passes through the magnetic coil. Because the adjustments in focal length may vary over a wide range (up to 10 cm) while simultaneous extreme precision in focusing is required, several controls of progressively increasing sensitivities are provided.

Either as part of the final lens, or positioned between the last two lenses, are deflection coils which move the electron beam in a raster pattern over the sample. The beam is deflected x number of times by fields controlled by the scan generators (Figure 1-1) and deflection yoke, respectively. The pattern of deflection is a rectangle or square. The scan generator is synchronized with the deflection coils in the microscope's CRTs, and produces sweep signals to the column deflection yoke. This synchronization among the scan generator, CRT, and deflection yoke results in a 1 : 1 correspondence between the position of the electron beam on the sample and the image observed on the CRT. Because of this time sequencing, there is great depth of field in SEM images (Figure 1-1).

The time-sequencing pattern is analogous to a television set, the difference between the two being the scan rate. Whereas a standard American television image is composed of 525 horizontal lines, a SEM image is variable from 100 to 100,000 lines. This flexible scan rate is essential for high-resolution SEM, because slow scan rates (*i.e.*, a larger number of lines) are required for sufficient electron-beam/specimen interactions; *i.e.*, the more electrons that strike the sample, the more data that may be collected. By using a slower scan on the CRT, the image will be of better resolution. The slowest scan rates (30, 60, or 120 sec/scan) are used for photography; medium scan rates (0.5, 1, or 2 sec/scan) are used for focusing; the TV scan rate is used for column alignment and sample localization.

Magnification is also a function of the illuminating system lenses. Magnification is the ratio of the size of the display area on the CRT to the distance the probe is scanned. Because changing magnification only involves changing the area scanned, in general it is unnecessary to refocus or recenter the specimen over a relatively large magnification range. In fact, it is normal operating procedure to increase magnification two steps, focus, then return to the desired lower magnification.

However, a parameter does exist here. To maintain the level of resolution from low to high magnifications, a smaller beam diameter should be used as magnification increases. Consequently, resolution is partially dependent on beam diameter. Because the smaller probe diameter results in lower electron emission per unit time, very long scan rates are used for high-resolution recording.

The final lens of the microscope is equipped with a movable aperture, which functions to remove noisy electrons. The noisy electrons may be either stray electrons from the column or primary beam, and/or electrons that have been backscattered from the sample. As a rule, the smaller apertures are used for imaging, because they afford the highest degree of contrast and a better signal-to-noise ratio.

Apertures are subject to a high degree of contamination and require cleaning after around 80 hr of use, or whenever astigmatism degrades resolution. Molybdenum apertures are cleaned in a high-vacuum bell jar system by heating to just below melting. Platinum apertures are chemically cleaned by boiling in concentrated sulphuric acid.

Most SEMs equipped with movable apertures require that the aperture be positioned in the center of the optical axis during normal operation. If the aperture is off-axis, resolution will be severely degraded. Apertures are aligned as follows:

1. Focus the image at approximately 500×. Observe the image movement going through focus. If the image does not expand uniformly from its center, proceed to step 2.
2. Adjust the *x* and *y* movements of the aperture until off-center movement is adjusted to expand from the center. Going through the focus clearly reveals the "best" focus, *i.e.*, when the aperture is centered.
3. Repeat steps 1 and 2 at approximately 1000×.
4. Check the electronic gun alignment (*x* and *y* axes) for the brightest image.
5. Adjust the stigmator to eliminate any distortion.

INFORMATION SYSTEM

The second system of the SEM consists of the sample, which will interact with the scanning electron beam and generate a variety of information signals. The SEM is capable of generating data in the form of three different electron signals (secondary, backscattered, and Auger electrons), specimen currents, photons, and characteristic x-rays. However, these signals are analyzed only if specific detectors–*e.g.*, a Si(Li) detector for x-rays–are present; otherwise, a typical SEM can only detect secondary electrons and some backscattered electrons.

Basically there are two types of interactions which generate an SEM electron image and may induce the other data signals: These are elastic and inelastic collisions. In an elastic collision, the incident probe electron is deflected by the sample without any energy loss–*i.e.*, the primary electron is now a backscattered electron (e_B^- or BSE). Inelastic scattering results in the loss of energy as the primary electron is absorbed by the sample, and its energy is transferred to atoms within the sample. To accommodate this gain in energy, the excited atom may then release an electron and return to the ground energy level (normal, unexcited state); the new electron released by the sample's atom is a secondary electron (e_2^-). Both e_B^- and e_2^- can contribute to image formation, although statistically the image is composed mainly of secondary electrons.

This becomes clearer when examining Figure 1-3: backscattered electrons tend to bounce back toward the source, thus BSE detectors are mounted beneath the final aperture. In comparison, the e_2^- detector is placed at a 90° angle relative to the primary beam. The sample is usually held at a 30°–60° angle relative to the beam, which means that both the electron signals will be directed toward the electron detector.

The type and relative number of electrons released from the sample depends to a large degree upon the atomic density, Z, of the site. Materials of high Z (*e.g.*, metals) will generate many backscattered electrons, whereas materials of low Z (*e.g.*, biological specimens) will absorb electrons and give rise to a larger secondary signal. That is, the depth of penetration and the degree of absorption by the probe is a function of the sample density. This is shown in Figure 1-3: a teardrop shape is characteristic of a specimen of low or medium atomic density, whereas a hemisphere is typical for a high-Z material.

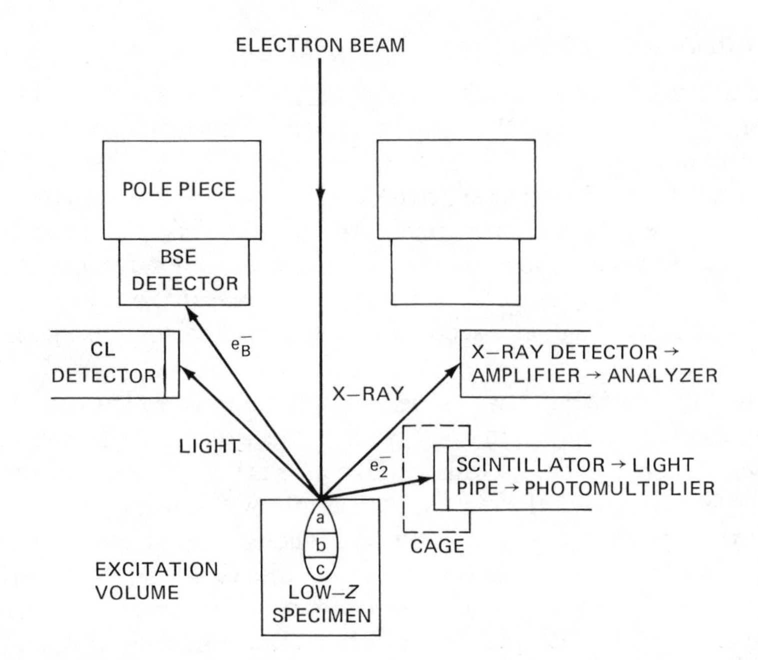

Figure 1-3. Data signal origin and detection. Excitation volume levels: a, secondary emission level; b, backscatter emission level; c, x-ray excitation volume. Cathodoluminescence and Auger electrons are emitted from approximately the top 10 Å of the excited surface.

Another factor controlling electron emission is the incident energy of the probe. Typical SEMs have a range of accelerating voltages (5–50 KeV); the choice of an optimal voltage depends to a large degree upon the type of specimen (although 15 KeV is typical). The accelerating potential influences the depth of penetration in the specimen, which in turn influences the strength of the emission signal.

Mathematically one relates Z to accelerating voltage by (Black, 1974):

$$d_p \propto - \frac{W_a V_o^2}{Z\rho}$$

where

d_p = depth of penetration
W_a = atomic weight

V_o = accelerating voltage
ρ = density

However, at low voltages or with very-rough-surfaced samples, it is the geometry of the specimen surface relative to the beam that has a greater influence on the intensity of the emission signal. These effects are interpreted as contrast in electron micrographs. For example, many samples are not smooth but rough surfaced. Therefore, a micrograph exposed in a given plane (tilt axis) will show certain areas of brightness/contrast; tilting the sample and re-exposing the same field of view will result in different areas of brightness/contrast. Differences in contrast result from depressed areas releasing a weak signal, whereas bright areas release a strong signal. (Contrast enhancement techniques, such as the use of gamma, are used to modulate the emission signal: this will be discussed later.)

Other factors influencing electron emission and closely related to Z are the surface chemistry and the effect of charge accumulation on the sample surface. It is these factors which are primarily responsible for the metal-coating techniques used to render naturally nonconducting materials (such as biological material) conductive. If a nonconducting sample is examined in its native state under normal operating conditions, electrons will accumulate in the area being examined and give rise to charging artifacts because the sample is not grounded. Consequently, this excited area will deflect the primary beam and severely degrade resolution.

In summary, five interrelated parameters control electron emission:

1. Specimen effects
   a. The atomic density Z of the specimen
   b. The surface chemistry and crystallography of the specimen
   c. Local charge accumulation
2. Instrumental effects
   a. Accelerating voltage
   b. Surface morphology and geometry of the primary beam *vs.* the sample *vs.* the detector

Another, less esoteric, factor easily controlled by the operator is working distance (WD), defined as the separation between the final lens aperture and the specimen surface. It is regulated by manipulating the *z* (vertical) axis. Optimum imaging occurs when the sample is as close to the lens as possible, provided that the specimen does not contact the lens. Working distance also influences magnification, in that as the distance between the specimen and lens increases, magnification decreases.

DETECTION SYSTEM

The detection system of the SEM is responsible for collecting and amplifying the information signals generated by the probe/specimen interactions. All microscopes have an electron detector situated 90° away from the beam axis. This orientation is best suited for secondary-electron detection; if a backscattered electron is deflected toward the detector, it will be detected also. A more sophisticated SEM will have a detector, especially for BSEs, mounted beneath the pole piece on the column—in addition to the normal detector. However, one must realize that a typical detector does, indeed, collect both electron signals; various manipulations are possible to distinguish one signal from the other.

Other detectors are used to collect the other emissions from the specimen. A monochromator-photomultiplier detects cathodoluminescence; photons are emitted in much the same way as x-rays from an excited atom, but are of lower energy. Lithium-drifted silicon detectors [Si(Li)] detectors-spectrometers analyze x-rays from the sample.

Secondary electrons released from the sample are attracted towards an electron collector held at a positive potential of 40–200 V with respect to ground. Because backscattered electrons are usually deflected back toward the column, they will not enter the collector unless, by chance, it is in their path. In this event, e_B^- may be distinguished from e_2^- by having zero potential across the electron collector: e_B^- are sufficiently energetic to travel away from the sample without additional positive attraction. However, e_B^- can generate e_2^- anywhere in the specimen chamber, and these noisy electrons may compose up to 30% of the final image (especially if the sample is of high density). In short, the CRT image is generated by two meaningful electron signals (data), plus noise (nonsense). The strength of the signal at the collector is proportional to the number of secondary electrons incident on it: The majority of signal detected here is composed of secondary electrons.

The electrons collected by the detector are then accelerated into a scintillator by a 10–12.5-keV field, and are translated into a proportional number of photons. The photons travel down a light pipe toward a photomultiplier, which in turn generates an amplified photocurrent signal used to modulate the CRT brightness (Figure 1-3).

The scintillator is an aluminum reflector surface, and literally serves as a mirror directing the photon signal. Because this aluminum surface is exposed in the specimen chamber and directed toward the sample, it is subject to extreme contamination caused by repeated exposure to the atmosphere and from sample sublimation during examination. Contamination will progressively degrade resolution, because it interferes with efficient signal collection. Therefore, for optimal SEM performance the scintillator must routinely be replaced, with about the same frequency that apertures are cleaned. They may be purchased at some expense from manufacturers; alternatively, the dirty aluminum can be removed and a fresh surface prepared by vacuum evaporation onto the disk. However, the thickness of the coating is critical, and a very clean vacuum system is essential. Some manufacturers require a film as thin as 50 Å, whereas others require 300 Å. The major difficulty, here, is in exactly determining thickness.

DISPLAY SYSTEM

The display system of the SEM serves to faithfully reproduce the topography of a specimen in the same sequence as it was electronically scanned. Synchronization exists between the scan generator and the deflection coils of the cathode-ray tube (CRT), and in this manner the scanning raster is reproduced on the CRT observation screen. The actual time sequencing has been previously discussed.

Usually the display system consists of two CRTs, one for observation and the other for image recording. The observation screen is coated with a fluorescent material (*e.g.*, zinc or cadmium sulfides) which reproduces the data from the photomultiplier. A Polaroid camera is then used to record the image.

Some older instruments use one CRT for observation and recording; two screens are better because:

1. Residual halation on the screen will introduce noise, resulting in hazy or too bright photographs.
2. The photographic CRT has a much finer fluorescent grain than the visual CRT, and thus can reproduce a more detailed signal.

VACUUM SYSTEM

The vacuum system of an electron microscope is designed to remove gases from the column for four reasons:

1. Gas molecules interact with high-velocity electrons and scatter them randomly, giving rise to glare, noise, or reduced contrast in the image.
2. Gas in the gun chamber gives rise to ionization and random electrical discharges, causing instability in the filament and flicker in the beam.
3. Residual gases react with the filament, eroding it.
4. Gas may contaminate the sample.

An electron microscope must be maintained at a pressure of 10^{-4} torr, or better, because the pathlength of an electron at that vacuum is 2.5 m: This is a major reason why transmission electron microscopes (TEMs) have columns about that length. The ideal method of operating an EM would be to evacuate the column, seal it, and switch off the vacuum pumps during observation. Unfortunately, this is impossible, because gases are continuously finding their way into the system through minute leaks and by sublimation of the sample during irradiation. Thus an EM is always held under vacuum, and contamination is minimized by leaving the vacuum system on continuously.

The degree of vacuum required depends primarily upon the type of electron gun, although, as a rule, the higher the vacuum, the better the resolution. Although a typical tungsten filament requires a minimum vacuum of 10^{-4} torr, most commerically available SEMs operate in the 10^{-5}-torr range. In comparison, a lanthanum hexaboride gun operates at least at 10^{-6} torr; and a field emission source at 10^{-10} torr at the gun level and 10^{-6} torr in the column.

The current state of vacuum technology requires that high vacuum be produced in two stages. The stages are arranged in series, each magnitude being produced by a different type of pump. The first stage is the creation of low vacuum, about 10^{-2} torr, from atmospheric pressure. Another pump then takes the system from low to high vacuum, or about 10^{-5} torr. Low vacuum is produced by a rotary mechanical pump (Figure 1-4), and high vacuum is achieved by a diffusion pump (Figure 1-5).

Rotary Pumps

The rotary pump (synonyms: mechanical, roughing, low vacuum pump) is an oil-immersed, eccentric-vane pump (Figure 1-4). Its construction is as follows. A cylindrical rotor, bearing along its length a pair of spring-loaded vanes, is mounted inside a cylindrical casing of slightly larger diameter. The rotor is eccentrically mounted so that it just touches the casing at one point, giving rise to a line contact. The spring-loaded vanes also bear on the casing via a pair of line contacts. The space between the rotor and casing is thus divided into

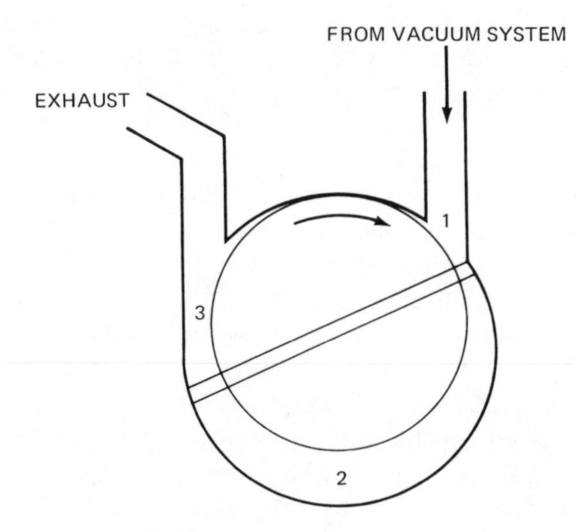

Figure 1-4. Cross-section of a rotary pump.

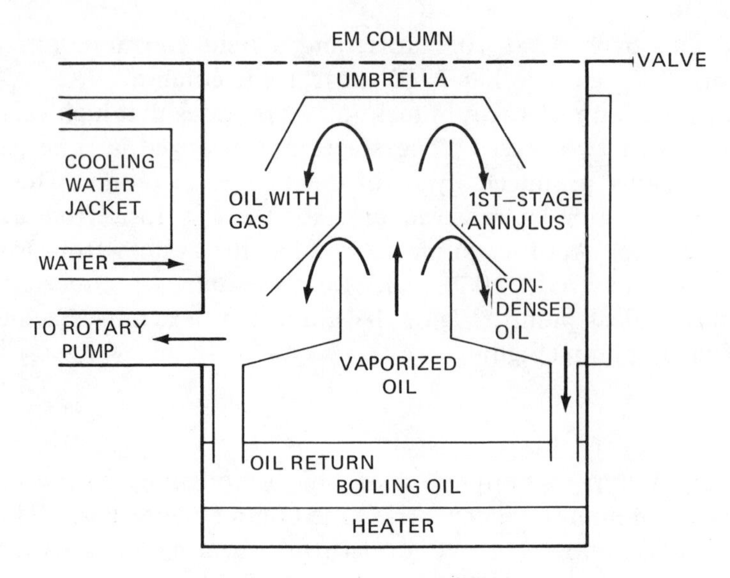

Figure 1-5. A two-stage oil diffusion pump.

three compartments by the two line contacts between the vanes and the casing, and the one line contact between the rotor and the casing. The spaces are numbered 1, 2, and 3 in Figure 1-4.

When the rotor revolves, space 1 becomes larger and draws gas from the microscope column, when going from atmosphere to low vacuum, or from the diffusion pump at high vacuum. Space 2 remains approximately the same size, but 3 becomes smaller. Any gas trapped in space 3 will be compressed and forced out the exhaust pipe to the atmosphere; however, as soon as the tip of the vane passes the orifice of the exhaust pipe, space 2 becomes space 3, and the gas trapped in space 2 is forced out. Simultaneously, space 3 becomes space 1, and the entire cycle is repeated. Thus, there is a continuous cycle of suction, idle, and exhaust.

The highest vacuum which a rotary pump can achieve during suction is dependent on the rate at which gas leaks across the vane seals and the rotor-to-casing seal. Consequently, rotary pumps are susceptible to pronounced mechanical wear which may severely decrease efficiency. They are also susceptible to overheating, because heating increases mechanical clearance by expansion. The highest vacuum to be expected from a single-stage rotary pump in optimal condition is 10^{-2} torr; when arranged in series some pumps achieve 10^{-4} torr.

The maintenance of a rotary pump is quite simple. Weekly, check the oil level at the indicator on the pump, and fill with the recommended oil. Yearly, replace the oil. Also check the motor's fan belt; periodically it will break. The fan belt must not be rubbing against any other part of the pump or any shields.

Rotary pumps may be located at any distance from the SEM; many are placed in separate rooms to reduce the annoying noise level. However, the more connections and/or distance between this pump and the SEM, the greater the chance of leaks developing. As a result, hose connectors should be checked if there is any problem with vacuum leaks.

Diffusion Pumps

The second pump, used to create high vacuum from the low vacuum produced by the rotary pump, is the diffusion pump. It contains no mechanical moving parts, making it something of an engineer's dream. This pump is mounted immediately beneath the sample chamber, making it highly efficient. The construction and operation of a diffusion pump is as follows (Figure 1-5). Oil or mercury is heated and vaporized at the base of the diffusion pump, causing it to move rapidly upward through the pump's center. As the vaporized oil emerges from the top of the tube the stream is deflected almost 180° by an umbrella. On its downward path, the oil or mercury vapor traps gas molecules and has sufficient impetus to "throw" gas molecules away from the microscope column. This action establishes a pressure difference across the pump umbrella, giving rise to the pumping action. Therefore, gas is free to attain equilibrium by diffusing from the column toward the diffusion pump, but it cannot diffuse back.

The actual pressure difference depends upon a number of factors, the primary one being the clearance between the umbrella and pump casing. The larger this annular area is made, the more rapidly will the pump reach its ultimate vacuum, and the higher will that vacuum be. However, if the gas pressure in the space below the umbrella rises above a critical value, then the gas which has been removed already will diffuse back into the column across the barrier of moving oil molecules, and the pump will cease to operate. Consequently, it is necessary to keep the space below the annulus continually pumped.

This is done by having a second stage umbrella mounted beneath the first. The annular space is smaller, so that the concentration of oil molecules is increased, and this stage of the pump can transfer gas from the intermediate pressure region to the space below. This process may be repeated a number of times, the annular space being reduced in area at each step. The limit is reached when the distance between the last umbrella and the casing has reached the reasonable limit of machining tolerance.

When the oil vapor has crossed the annular space of each stage, it strikes the water-cooled pump casing. The vapor then condenses to a liquid and returns by gravity to the boiler, where it is vaporized and recycled. The gas which accumulates below the final stage must be continuously removed by connection to the rotary pump.

In a diffusion pump using mineral oil as the pumping fluid, each stage deals with a pressure difference of about a factor of 10 across it. Four stages will therefore give a total pressure-difference factor of 10^4—or a high vacuum of 10^{-5} torr, if the backing (low) vacuum pulls 10^{-1} torr. The highest vacuum is determined by the vapor pressure of the pumping fluid at the temperature of the walls of the condenser. With the oils commonly used, this is in the 10^{-6} torr range.

A number of precautions must be observed when operating a high-vacuum system. First, never open the valve of a hot diffusion pump to the atmosphere. Although modern SEMs have a failsafe system which prevents this, older microscopes do not, nor do vacuum bell jar systems (see Chapter 9). If this happens, the microscope or bell jar will be filled with burned oil, more familiar as asphalt. It is unpleasant to remove this deposit from a vacuum system. Second, always make sure that there is sufficient water flow to cool the pump. Overheating the oil, unless it is silicon-based, rapidly contaminates it.

In summary, a high vacuum operates as follows:

1. Low vacuum is achieved from atmospheric pressure with a rotary pump. Thus, there is a valve between the rotary pump and SEM.
2. After low vacuum is achieved, open the warm diffusion pump to the column, isolating the rotary pump from the column, and open the valve between the rotary pump and the diffusion pump.

3. To return to atmosphere:
   a. Close the valve between the diffusion pump and the SEM.
   b. Admit air.
   c. Let the rotary pump continue to evacuate the diffusion pump.

Vacuum Indicators

SEMs are equipped with "idiot lights" to indicate when high vacuum is reached. However, direct monitors of both low and high vacuum are desirable as aids in the detection and isolation of vacuum leaks. Low vacuum is monitored with a Pirani gauge, and high vacuum with an ionization gauge.

In a Pirani gauge, a hot filament is cooled by convection–*i.e.*, by gas molecules coming into contact with the filament–which carries off thermal energy at a rate dependent upon the number of molecules present. The cooling effect is determined by observing the resistance of the filament.

In some ionization gauges, a rarefied gas is subjected to ionization by thermionic emission from a heated filament; another electrode, held at lower potential than the filament, then carries off the positive ions formed. The tube is designed such that the ionization current is proportional to the gas pressure.

RESOLUTION

The factor which makes electron microscopes powerful tools is their high resolution. Resolution is defined as that distance between two adjacent objects when the objects lose their separate identities. Simple magnification beyond this point is worthless, since one is enlarging an inherently blurred image.

Discussed below are three distinct but interrelated groups of factors which determine resolution (Black, 1974). The first group is concerned with electron optical performance, i.e., instrumental factors governed by the laws of physics. These limitations are inherent in optical systems, and in turn are responsible for the maximum possible resolution. The second is concerned with practical interference problems, such as contamination and the influence of stray electrical and magnetic fields. The third includes interactions between the beam and specimen, such as the amount of information that may be learned and the degree of damage caused by vacuum exposure and irradiation. The first set of parameters are beyond operator control, but the last two may be manipulated.

Electron Optical Limitations

Theoretically, the resolving power of the optical microscope is limited by the de Broglie wavelength of light, and that of the electron microscope is ultimately controlled by the wavelength of electrons. High resolution requires a small ratio of wavelength to lens diameter: a light microscope uses visible light having a wavelength of several thousand Angstroms, whereas an electron microscope's beam is of wavelength less than 1 Å. However, there are four physical factors which severely limit resolution, i.e., although wavelength ultimately defines resolution, the construction of instruments cannot overcome other optical phenomena. They are diffraction, astigmatism, and chromatic and spherical aberrations.

Diffraction is the interference between the component rays of a single broad wave front (Fig. 1-6). Visibly it occurs when light passes near an object where the shadow would be. Diffraction determines the maximum resolution obtainable with any type of microscope. This concept was mathematically proven to be related to wavelength and thus to resolution.

In the nineteenth century, Abbe defined the magnitude of diffraction effects as d, where d provides a numerical value for the limit of

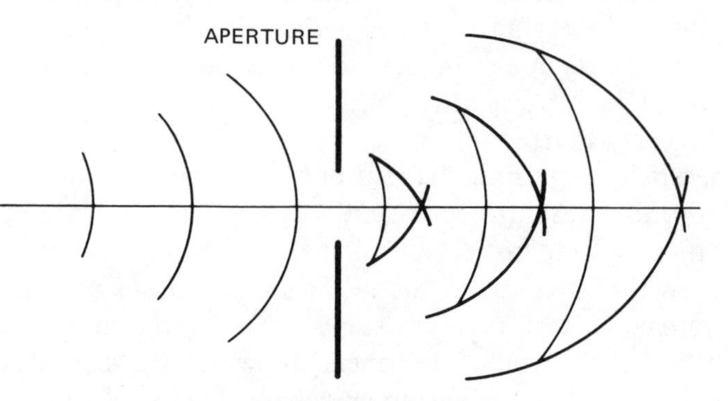

Figure 1-6. Diffraction effects.

resolution of any diffraction-free optical system:

$$d = \frac{0.612\lambda}{n \sin \alpha}$$

where

d = radius of the first dark ring of an Airy disk measured at the minimum.
λ = wavelength of the image-forming radiation.
n = index of refraction between the point source and the lens, relative to free space.
α = half the angle of the cone of radiation from the specimen plane accepted by the front lens of the objective.

Most light microscopists are familiar with the relation

$$n = \sin \alpha$$

which is usually referred to as the numerical aperture of a lens.

From the Abbe equation, it can be derived that to attain maximum resolution with a light microscope, it is necessary to have a maximum n and maximum $\sin \alpha$. Because n cannot be increased beyond approximately 1.5 and α near 90°, the only remaining factor to enhance resolution is to decrease λ. As mentioned above, visible light has a wavelength of about 2000 Å, whereas the wavelength of electrons is less than 1 Å.

de Broglie advanced the idea that moving particles have wavelike properties. Implicit in this theory is the fact that an electron beam can be used as a type of illumination. de Broglie then concluded that a wavelength can be assigned to moving particles, which may be calculated from

$$\lambda = \frac{h}{mv}$$

where

λ = wavelength of the particle

h = Planck's constant
   $= 6.23 \times 10^{-27}$ erg-sec
v = velocity

When the known values for electrons are substituted, it becomes

$$\lambda = \frac{12.3}{V} \text{Å}$$

where V = accelerating voltage.

Thus the wavelength (and resolution) of a beam of electrons depends upon the potential, V, through which is has been accelerated.

Abbe's equation can then be rewritten, substituting for the de Broglie equation for λ:

$$d = \frac{(0.61)(12.3)}{n \sin \alpha \sqrt{V}}$$

Because electron microscope angles are always very small, $\sin \alpha \doteq \alpha$, and since both the object and image are in field-free space, the refractive index is n = 1. The above equation is then rewritten as:

$$d = \frac{7.5}{\alpha \sqrt{V}} \text{Å}$$

Therefore, resolution in an electron microscope is ultimately determined by the accelerating voltage and angular aperture of the objective lens. Substituting realistic values in the above equation (voltage = $10^5 v$), d = 2.4 Å. The resolution of the light microscope is 2000 Å. This magnitude in difference is responsible for the greater resolving power of the electron microscope.

It must be stressed that the theoretical limitation of resolution as a result of diffraction effects is much better than practical values. Under optimal conditions, typical SEM resolution is 50–70 Å, while routine operation using a well-prepared sample yields resolution of 70–100 Å.

Diffraction in the TEM is manifested as a spreading of the beam behind the sample. This is not the electron diffraction, which is used

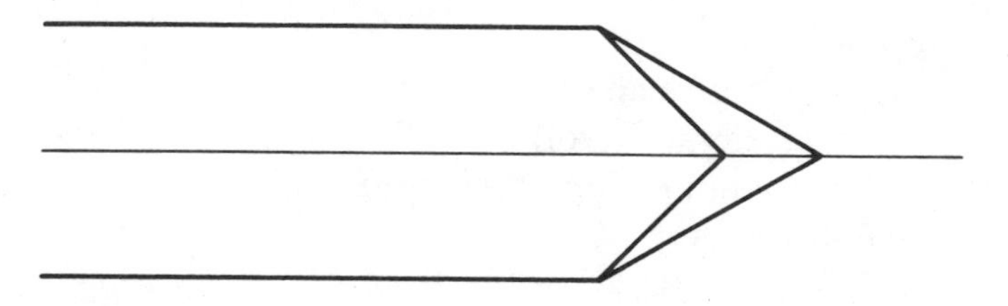

Figure 1-7. Chromatic aberration.

to identify compounds or elements on the basis of their crystalline nature. Diffraction effects in SEM are apparent in crystals of low Z. Here, the electron beam penetrates the sample and spreads, resulting in electron noise. This problem can be reduced by coating the sample by the sputter or evaporative technique.

Chromatic aberration is another parameter affecting resolution (Fig. 1-7). It arises because the focal length of a lens will vary according to the wavelength of radiation passing through it. In the light microscope, it is manifested by light of different colors being focused at different distances from the lens; it has been overcome by using flint glass lenses in conjunction with typical glass lenses.

In the electron microscope, chromatic aberration involves electrons travelling at slightly different velocities (and therefore different wavelengths) due to fluctuations in the accelerating voltage. Consequently, rapidly moving particles pass through a lens' magnetic field quickly without too much influence and are focused further from the lens than electrons traveling at more moderate speeds.

Mathematically, chromatic aberration is related to resolution by

$$d_{cV} = k_c \cdot f \cdot \alpha_o \cdot \frac{\Delta V}{V}$$

$$d_{cI} = 2k_c \cdot f \cdot \alpha_o \frac{\Delta I}{I}$$

where

d_{cV} and d_{cI} = separation of two object points which are just resolved, considering voltage and current respectively.

k_c = dimensionless constant (0.75 - 0.1)

f = focal length
α_o = objective aperture angle
V = accelerating voltage
ΔV = maximum departure from V
I = current
ΔI = maximum departure from I

Chromatic aberration is largely eliminated by incorporating a very stable high-voltage system in the electron microscope (see below), but it is never completely eliminated because of electron-specimen interaction.

Rewriting the above equation will demonstrate the maximum fluctuation tolerable in a high-voltage system:

$$\frac{\Delta V}{V} = \frac{d_{cV}}{k_c \cdot f \cdot \alpha}$$

Substituting realistic values,

d_{cV} = 10 Å; the maximum resolution resulting from diffraction and spherical aberration
k_c = 0.75
$\alpha = 4.5 \times 10^{-3}$ radians (optimum aperture angle)
f = 3 mm

Therefore,

$$\frac{\Delta V}{V} = \frac{10 \text{ Å}}{(0.75)(4.5 \times 10^{-3} \text{ radians})(3 \text{ mm})}$$

$$= 1.0 \times 10^{-4} \text{ of the voltage.}$$

One may also calculate the variation in lens current as related to chromatic aberration by

$$d_{cI} = 2k_c \cdot f \cdot \alpha \frac{\Delta I}{I}$$

$$\frac{\Delta I}{I} = 5 \times 10^{-5} \text{ of the current.}$$

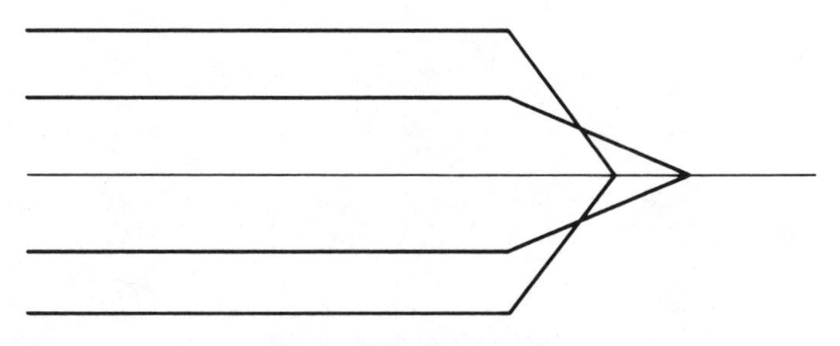

Figure 1-8. Spherical aberration.

The third optical parameter influencing resolution is spherical aberration, which depends upon the aperture angle of the incident radiation (Fig. 1-8). It is the most serious geometric aberration encountered in electron optics. It occurs when electrons leaving the object and passing close to the center of the objective lens are focused in one plane, whereas electrons passing through the outer edge of the magnetic field are focused in another plane. Consequently, electrons passing through the periphery have a short focal length, because the magnetic field is always much stronger at the perimeter.

The only way to reduce spherical aberration is to introduce an aperture just beneath the final lens. It allows only those electrons passing through the optical axis to form the final image; peripheral electrons are stopped by the aperture. Unfortunately, the aperture severely reduces resolution because it limits the numerical aperture of the lens. That is,

$$d_s = k_s \cdot f \cdot \alpha_o^3$$

where

d_s = separation of two object points which are just resolved.
k_s = dimensionless proportionality constant.
f = focal length.
α_o = objective aperture angle.

The final optical limitation influencing resolution is astigmatism, which develops when a lens focuses more strongly along one axis than along another due to minute flaws or inhomogeneieties within the coilings (Fig. 1-9). As a result, the image can never be clearly

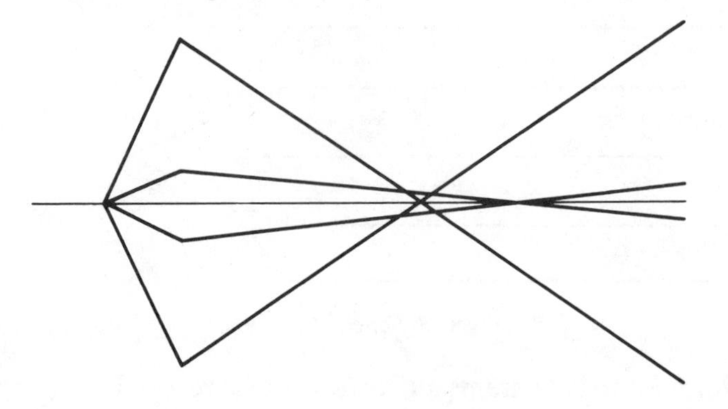

Figure 1-9. Astigmatism.

focused. It is corrected by superimposing on the objective lens magnetic field another field having a deliberate asymmetric distribution and variable magnitude; it is then positioned so as to oppose and cancel the existing lens' asymmetry. The correction is made by a stigmator, which functions as an additional weak cylindrical lens of strength just sufficient to correct the cylindrical component of the objective.

Astigmatism is manifested by a change in shape, for example, of a round object into an oval, as focus is varied. Unless it is extremely pronounced, it becomes apparent in the SEM at magnification greater than 2500 X. Therefore, it should be corrected at high magnification by "focusing" with the stigmator until the phenomenon is eliminated. An image is not stigmatic when a shape remains the same as focus is varied. That is, a circle symmetrically expands out of focus and back into focus; it does not go from a circle into an oval as focus is altered. All samples must be checked for astigmatism.

In summary, diffraction and spherical aberration effects are beyond operator control, whereas chromatic aberration and astigmatism are influenced by proper operation.

There are a number of additional optical parameters associated with the electron microscope. These are beam brightness, accelerating voltage, and lens operating conditions. Although these subjects have been touched upon previously, the following discussion serves to correlate that information.

The characteristics of standard electron guns have been discussed. The major distinction among these sources is beam brightness as a function of beam diameter at the sample. Beam diameter d is (Black, 1974):

$$d^2 = d_o^2 + d_s^2 + d_c^2 + d_f^2$$

where

d_o = aberrationless beam diameter at specimen.
d_s = diameter of disk of confusion due to spherical aberration.
$= \frac{1}{2} C_s \alpha^3$, where C_s is the spherical aberration constant.
d_c = diameter of disk of confusion due to chromatic aberration.
$= (\Delta V/V)(C_c \alpha)$, where C_c is the chromatic aberration constant, V = accelerating voltage, and $\Delta V = V$ instability.
d_f = diffraction effect or the Airy disk diameter, where

$$d_f = 2r = \frac{1.22\,\lambda}{\alpha}$$

The final beam current i is (Black, 1974):

$$i = \frac{\pi}{6} d_o^2\, J_c \frac{eV}{kT} \alpha^2$$

where

J_c = emission in amp/cm^2 at the gun (i.e., beam current density).
T = absolute temperature in °K
k = Boltzmann's constant
α = aperture angle
V = accelerating voltage
e = electron charge

The beam current density at the image plane J_i can then be calculated from (Black, 1974):

$$J_i \simeq J_c \frac{eV}{kT} \alpha^2$$

To improve the number of electrons incident on the specimen, and therefore the signal-to-noise ratio,

1. Increase the accelerating voltage.
2. Increase the objective aperture angle or decrease the working distance.

3. Use a different electron source
   1) LaB_6 will emit at lower temperatures.
   2) LaB_6 and FE sources have a higher specific emission than pointed tungsten.

To summarize the above equations and solve for beam diameter,

$$d_o^2 = \frac{6i}{\pi^2 \beta} \alpha^{-2}$$

where

$$\beta = \frac{J_c eV}{\pi kT}$$

or

$$d_o = C_o/\alpha$$

where

$$C_o = \sqrt{\frac{6i}{\pi^2 \beta}}$$

which can be reduced to

$$d^2 = \frac{C_o^2}{\alpha^2} + \frac{1.5\,\lambda^2}{\alpha^2} + C_s^2 \alpha^6 + \left(\frac{C_c \Delta V \alpha}{V}\right)^2$$

Substituting realistic values for the SEM operating at 20 keV,

$Cs = Cc = 1$ cm

$i = 10^{-11}$ amp

$\beta = 10^5$ amp/cm^2/ster

$$\frac{\Delta V}{V} = 10^{-4}$$

$$d_o = d_c = 50 \text{ Å and intersect at } \alpha = 0.005 \text{ radian}$$

$$d = \sqrt{50^2 + 50^2}$$

$$= 70 \text{ Å}$$

which is the typical resolution of the SEM. A more intense gun reduces the value of d_o and improves resolution to 30 Å. The primary optical effects discussed above usually raise these values.

Practical Interference Problems

This group includes mechanical vibrations, stray electrical or magnetic fields, and specimen contamination. Whereas optical limitations cannot be controlled by the operator, practical interference problems may be readily dealt with and eliminated.

Prior to installment, an engineer will inspect the proposed site for any stray electrical or magnetic fields. Direct power supplies will emit stray fields which will interfere with the electron microscope lenses. In most cases, microscopes are situated where extraneous noise is minimal.

Vibrations are also a problem in electron microscopes, but are controlled in the same manner as extraneous fields. They can be a severe problem in desktop SEMs, where vibrations are transmitted through the desk to the column. Because it is usually impossible to eliminate the source of vibration, the microscope column is mounted on an isolation pad. Other electron microscopes are equipped with floating columns, in that large air bags absorb noise. Vibrations are visible as scalloped edges on what would be a smooth-edged sample.

Specimen contamination is a particularly bothersome problem, but it is again within operator control. Contamination arises from sublimation of the sample, either by the vacuum (*e.g.*, an acetone-cleaned sample may have residual oil or acetone on it—let it dry before putting it into the electron microscope!) or as a result of radiation damage (low Z materials will burn away). Contaminants will be deposited on all internal surfaces of the column; the cleaning of these surfaces has been discussed.

Beam-Specimen Interactions

The final group of factors which influence resolution are beam–specimen interactions, which can be broken down into beam penetration and scattering, electron yield, specimen composition and geometry, and specimen damage from secondary radiation effects (*e.g.*, heating). These factors have been discussed previously and will not be repeated.

REFERENCES

Ardenne, M. von (1938). The scanning electron microscope: practical considerations. *J. Phys.* **109**:533.

Barbi, N. G. (1980). Detectability in energy-dispersive microanalysis. *SEM, Inc.* 2:297.

Black, J. T. (1974). The scanning electron microscope: operating principles. In: *Principles and Techniques of Scanning Electron Microscopy* 1:1. (M. A. Hayat, ed.) Van Nostrand Reinhold, New York.

Boyde, A. (1979). The perception and measurement of depth in the SEM. *SEM, Inc.* 2:67.

Chandler, J. A. (1979). Principles of x-ray analysis in biology. *SEM, Inc.* 2:595.

Crewe, A. V. (1971). A high-resolution scanning electron microscope. *Sci. Amer.* **224**:26.

Erasmus, D. A., ed. (1978). *Electron Probe Microanalysis in Biology*. Chapman and Hall, London.

Everhart, T. E. and T. L. Hayes (1972). The scanning electron microscope. *Sci. Amer.* **225**:54.

Fujita, T. *et al.* (1971). *Atlas of Scanning Electron Microscopy in Medicine.* Igaku Shoin Ltd., Tokyo.

Goldstein, J. I. and H. Yakowitz, eds. (1975). *Practical Scanning Electron Microscopy*. Plenum Press, New York.

Greer, R. T. (1976). Fundamentals of SEM for physicists. *IITRI/SEM* **1**:669.

Grivet, P. (1972). *Electron Optics*, 2nd ed. Pergamon Press, New York.

Hayes, T. L. (1980). Keynote paper: Biophysical aspects of scanning electron microscopy. *SEM, Inc.* **1**:1.

Holt, D. B. *et al.*, eds. (1974). *Quantitative Scanning Electron Microscopy*. Academic Press, New York.

Hren, J. J. *et al.*, eds. (1979). *Introduction to Analytical Electron Microscopy*. Plenum Press, New York.

Jones, A. V. and B. M. Unitt (1980). Computers in scanning microscopy. *SEM, Inc.* **1**:113.

Joy, D. C. (1977). Scanning electron microscopy (SEM)–where next? *IITRI/SEM* **1**:1.

—— and C. M. Maruszewski (1978). The physics of the SEM for biologists. *SEM, Inc.* **2**:379.

Kessel, R. G. and C. Y. Shih (1974). *Scanning Electron Microscopy in Biology–A Students Atlas on Biological Organization*. Springer-Verlag, New York.

Kirz, J. (1980). Specimen damage considerations in biological microprobe analysis. *SEM, Inc.* **2**:239.

Ludwig, H. and H. Metzger (1976). *The Human Female Reproductive Tract–A Scanning Electron Microscope Atlas*. Springer-Verlag, New York.

Muir, M. D. (1974). Fundamentals of the scanning electron microscope for biologists. *IITRI/SEM* p. 1011.

Newburg, D. E. (1977). Fundamentals of scanning electron microscopy for physicist: contrast mechanism. *IITRI/SEM* **1**:553.

Nixon, W. C., editor (1974). *Scanning Electron Microscopy Systems and Applications*. Institute of Physics, Bristol.

Oatley, C. (1972). *The Scanning Electron Microscope*. Cambridge University Press, Cambridge.

Pfefferkorn, G. (1975). Introduction to scanning electron microscopy. *IITRI/SEM* p. 631.

——(1977). A bibliography on fundamentals of SEM for biologists. *IITRI/SEM* **1**:569.

—— *et al.* (1978). How to get the best from your SEM. *SEM, Inc.* **1**:1.

Reimer, L. and G. Pfefferkorn (1977). *Rasterelektronenmikroskopie*, 2nd ed. Springer, New York.

Revel, J. P. (1978). Biological scanning electron microscopy for physicists and engineers. *SEM, Inc.* **1**:829.

Rochow, T. G. and E. G. Rochow (1978). *An Introduction to Microscopy by Means of Light, Electrons, X-rays, and Ultrasound.* Plenum, New York.

Roomans, G. M. (1980). Problems in quantitative x-ray microanalysis of biological specimens. *SEM, Inc.* **2**:309.

Seiler, H. (1976). Determination of the "information depth" in the SEM. *IITRI/SEM* **1**:9.

Spurr, A. R. (1980). Applications of x-ray microanalysis in botany. *SEM, Inc.* **2**:535.

Troughton, J. and L. A. Donaldson (1972). *Probing Plant Structure. A Scanning Electron Microscope Study of Some Anatomical Features in Plants and The Relationship of These Structures to Physiological Processes*. Chapman and Hall, Ltd., London.

Verhoeven, J. D. and E. D. Gibson (1976). Evaluation of a LaB_6 cathode electron gun. *J. Phys. E.* **9**:65.

—— and E. D. Gibson (1977). On the design of the Broers type LaB_6 gun. *IITRI/SEM* **1**:9.

Wells, O. C. (1972). Bibliography on the SEM. *IITRI/SEM* p. 375.

——, ed. (1974). *Scanning Electron Microscopy*. McGraw Hill, New York.

Welter, L. M. (1975). Application of a field emission source to SEM. In: *Principles and Techniques of Scanning Electron Microscopy*, Vol. 3, p. 195. (M. A. Hayat, ed.) Van Nostrand Reinhold, New York.

Yoshii, Z., *et al.* (1976). *Atlas of Scanning Electron Microscopy in Microbiology*. Igaku Shoin, Ltd., Tokyo.

Young, J. (1973). Update of bibliography on the scanning electron microscope. *IITRI/SEM* p. 775.

2. SEM Photography

INTRODUCTION

The image obtained on the fluorescent CRT screen of an SEM is both impermanent and of relatively poor resolution. The most convenient and inexpensive method to record images permanently is photography. The end-product is properly referred to as a scanning electron micrograph. Because most SEMs are equipped with Polaroid cameras, very little effort is expended in recording an image. However, producing a good micrograph requires some practice; this is the primary goal. Therefore, emulsions and image formation will be discussed in detail.

Although a Polaroid camera user need not be concerned with the following steps, the normal photographic system may be broken down into three steps:

1. *Exposure*: The image-forming radiation is allowed to strike the photographic emulsion for a suitable length of time.
2. *Development*: The latent image resulting from exposure is chemically developed, and the resulting transparency (negative) is fixed, washed, and dried.
3. *Printing*: The film of the negative image is placed in an optical enlarger, and a black-and-white print is exposed, developed, fixed, washed, and dried—yielding a scanning electron micrograph.

Polaroid film does all of the above for the user, but good image recording requires that the user understand these steps.

EXPOSURE

An ordinary black-and-white emulsion layer of photographic materials is basically a suspension of small crystals of silver halide in gelatin. The silver halide crystals are referred to as grains. The emulsion layer is 12–25 μm thick and consists of 10% silver halide dispersed as discrete grains, 0.5–10 μm in diameter, in gelatin. The gelatin protects the underlying grains against abrasion which would render them developable without exposure. Subsequent chemical treatment hardens the gelatin and protects the image.

The conversion of the electron signal into a light signal which is reproduced on the CRT screen has been discussed in Chapter 1. The phosphor on the CRT releases its excitation energy as quanta of light, which, when allowed to strike an emulsion, transfer that energy to a trap on a silver halide crystal. Aggregates of silver atoms are formed by combining the activated silver halide crystals and silver ions. These aggregates are referred to as latent-image specks: They may be formed at the surface of, as well as within, an emulsion grain. In general, commercial developers react with the latent-image speck on the grain surface.

The quantum efficiency of an emulsion with respect to light is relatively low: A single silver halide grain must be hit by 10–100 photons before it can be developed. Sensitivity to light varies as the cube of grain diameter. Therefore, the larger the grain, the more photons it will collect, and the faster will the emulsion react to light. Unfortunately, excessively large grains degrade resolution. This is offset in SEM by altering the scan rate: The slower the scan rate; the larger the number of incident electrons and electron signals; and the larger the number of phosphor crystals excited in the CRT–the more silver halide grains are activated in the emulsion.

A variety of emulsion parameters influence the efficiency of photographic exposure. In general, these are beyond operator control, but assist in interpreting micrographs.

Ratio of silver halide to gelatin: The most efficient utilization of light energy occurs if all of the light strikes a silver halide grain, thus rendering it developable.

Stopping-power of the silver halide: The number of grains struck will increase with decreasing stopping-power. In practice, the differences in halide composition among commercial materials available for SEM are such that stopping-power variability is negligible.

Degree of chemical sensitization: Sensitization consists of one or more treatments of emulsion grains, which reduce the exposure required to make a grain developable by

1. Increasing the efficiency of silver formation.
2. Reducing the dispersal of silver formed by exposure.
3. Reducing the quantity of silver required for developability.
4. Any combination of the above.

In general, as sensitization is improved, the number of grains rendered developable among those hit will increase.

Spread function: a measure of the average projected area over which radiated energy is expended. In practice, an emulsion is capable of resolving two discrete light beams if they are separated by a distance greater than the spread function of the emulsion. In general the width of a spread function is slightly less than the depth of penetration by the incident radiation. This phenomenon is independent of the grain size, but dependent upon the thickness and gelatin content of the emulsion.

Resolving power: the ability of an emulsion to record fine detail. If all fluctuations in the image-forming radiation could be eliminated, the limit of information storage in a material would be set by its spread function. Because photons are not massive, spread is minimal. An analogous situation, although exaggerated, is the spreading of a light beam in a turbid solution.

Signal-to-noise ratio of the photographic material: The signal is an optical density, and the noise is the fluctuation in density from area to area. If the photographic material records all of the incident radiation and does not introduce any noise of its own, it performs perfectly, and the signal-to-noise ratio of the photographic material is identical to that of the original CRT image. However, the information signal also has a signal-to-noise ratio, which will be discussed. It is useful to assess the degree to which any photographic material approaches this perfection: For this purpose, a concept known as the detective quantum efficiency has been developed.

Detective quantum efficiency (*DQE*): This is the assessment of the degree to which a recording device approaches perfection. It is defined as

$$DQE = \frac{(\text{signal})^2/\text{noise}_{\text{photograph}}}{(\text{signal})^2/\text{noise}_{\text{incident radiation}}}$$

A DQE value of 1 indicates a perfect recorder. $DQE = 0.25$ means that the signal-to-noise ratio is reduced by half in the recording step, or

$$DQE = \frac{(0.5)^2}{(1.0)^2} = \frac{0.25}{1.0} = 0.25$$

Graininess: Whenever a photographic image is magnified, it does not appear homogeneous because of the presence of silver halide grains. In talking about these grains, the term "graininess" is often introduced, and erroneously regarded, as a defect. It is not a defect, but simply a record of statistically random fluctuations, or noise, in the electron beam. Graininess is a statistical phenomenon that is several orders larger than the size of the grains that form the image. This misleading term has been discribed as "the impression of non-uniformity in the image which is produced on the consciousness of an observer by the granular structure."

Signal-to-noise ratio of the incident radiation: a measure of the image quality of the beam/CRT system. A photograph is only as good as the radiation used to make the exposure. Thus, the SNR of the incident radiation is very different from the emulsion limitations, above, and is the primary practical parameter when recording a CRT image: The SNR of the incident radiation must be modulated whenever an image is being recorded. A good electron micrograph extracts the meaningful data from a noisy background; thus, one wishes to minimize noise. In a routine situation, this is accomplished by collecting more data from the sample. Here, SNR is proportional to the square root of the number of electrons incident on the specimen. Consequently, slow scan rates increase the information signal by allowing more information to be released from the sample; and apertures and a small spot size reduce the number of noisy background electrons.

Many SEM users mistakenly believe that the SNR is reduced by increasing the development time and/or activity. This is not true, because signal and noise increase proportionately. Thus, the gain in contrast absolutely does not provide a method for distinguishing a weak data signal from noise.

Sophisticated techniques, such as optical filtering, also improve the SNR, but these are artificial means that are based on a photograph already having the best SNR for the specimen. Do not be misled by image enhancement techniques: They presuppose that all instrumental effects are optimal.

Contrast in SEM photographs is a variable characteristic which is difficult to evaluate. Defined as the ratio of light to dark areas in a photograph, contrast is more an aesthetic quality than a measurable quantity. A good micrograph is a range of gray tones rather than strictly black and white–*i.e.*, a range of tones which with minimal noise faithfully reproduces the topography of a specimen. This is analogous to adjusting brightness on a television set: Gray tones permit observation across the entire screen, rather than bright images superimposed on a black void. Contrast in the SEM is varied by use of contrast and brightness controls, which simply affect the viewing screen. These controls are not independent of each other; clearly, by increasing contrast (C), brightness (B) decreases. Contrast and brightness controls are simply electronic modulations of the signal. Most SEMs will recommend a numerical level for these controls, but they are only a starting point; *i.e.*, a C:B level of 6:4 is recommended for the Zeiss Novascan, but some samples may require a level of 7:3. Practical experience with photography typically solves these problems.

Because contrast/brightness controls the average contrast across the screen, but not for a given area, another electronic modulation of the image signal is available. The gamma control is used when details of the specimen appear washed out due to high brightness (*e.g.*, edges or crests on the sample) or lost in darkness (holes). Thus, one may bring the amplifier gain from normal to a point at which it is possible to see the bottom of a hole. In effect, the gamma control nonlinearly compresses or expands the video signal to delineate surface features without losing visibility in other areas of the display. When gamma = 1, the gain in contrast and brightness is identical; *i.e.*, the signal is amplified linearly: This is typically used for routine, rather smooth, samples. In comparison, when gamma = 0.5, the contrast gain is half as strong as brightness; *i.e.*, the contrast components in the dark region are strongly amplified, while those in the bright regions are only slightly amplified. Alternatively, by increasing gamma to 2.0, the contrast gain is twice as strong as the brightness gain—or, the contrast components in the bright regions are strongly amplified, while the components in the dark region are only slightly amplified. A low gamma should be used for rough surfaces and a high gamma for very smooth surfaces.

In summary, the following parameters directly affect the quality of a micrograph:

1. *Scan rate*: The highest resolution is obtained with the slowest scan rate.
2. *Objective aperture*: The smallest aligned aperture results in the least noise.
3. *Spot size*: Keep small.
4. *Contrast/brightness level*: Adjust for each sample.
5. *Gamma control*: Adjust for each sample.
6. *Tilt*: Determine on the visual CRT the optimum angle for observation.
7. *Focus*: Always focus in the visual scan rate two magnification steps higher than the final magnification desired.

TYPES OF FILM

Bascially, an SEM image is photographed by exposing Polaroid Land film and using a camera back for support; or on 35-mm film, using a conventional camera with special lenses.* Whereas the former is standard equipment on most SEMs, the 35-mm attachment is "special" optional equipment. The Polaroid system is used because it has the advantage of yielding micrographs immediately—*i.e.*, development by wet chemistry is unnecessary. On the other hand, a fairly wide range of 35-mm films is available, meaning that one may control to a larger degree film speed and, thus, resolution. The latter

*[The mention of specific manufacturers' products does not constitute an endorsement of these products.]

films also allow enlargement up to 10X, whereas Polaroid films cannot be magnified beyond about 4X. Because Polaroid films are most common, the following discussion is limited to that type of film. Recordings using a 35-mm camera are thoroughly discussed in Bertaud *et al.* (1978).

Two types of Polaroid film are typically used in SEM. Type 52 film yields a positive print, and Type 55 yields both a positive print and a negative. Both films are panchromatic, but differ in their speed and resolution. Consequently, it is necessary to adjust the brightness and contrast levels on the recording CRT, not on the observation CRT, to accommodate these differences when switching from one film type to the other. Type 52 film has an ASA of 400 and DIN of 27; in comparison Type 55 film has values of 50 and 18, respectively. In addition, the resolution of Type 52 film is 35–40 lines/mm, whereas that of the Type 55 print is 22–25 lines/mm and Type 55 negative is 150–160 lines/mm. The increased resolution of the negative permits enlargement with less loss of resolution in the final print.

Following exposure, the film is developed in its packet: in 15–20 sec for Type 52 or 20–25 sec for Type 55, at room temperature. When pulling the film from the camera back, do so with an even, continuous motion. Yanking the film out does not permit developer to evenly coat the film, whereas pulling too slowly results in uneven development. The prints must be coated as soon as possible with the coaters supplied with the film. The coating will prevent scratching and fading of the image; always stroke five or six times completely across the image and borders. Do not use the coaters for more than five prints. Allow the prints to dry at least 15 min to prevent damage. When labelling the backside of prints, never write over the emulsion area, only on the border.

The negative obtained with Type 55 film requires further treatment before printing. After separation from the print, the negative must be handled carefully because it is quite soft. Some developer will adhere to it, and must be immediately removed by submerging in 18% sodium sulfite with gentle agitation. If the sodium sulfite bath cannot be immediately reached, the negatives may be safely stored in water for about 3 min. However, the only sure way to prevent staining of the negative is to immediately place it in the sodium sulfite.

After agitation for about a minute in the sodium sulfite, the developer film comes off and the dye clears. For optimal enlargement, the negative is treated in an acid hardening bath for 1–2 min, then washed in cool running water for 5–15 min. The negative is then dipped in a wetting agent (*e.g.*, Kodak Photo-Flo) and air-dried. It is essential that all of the above solutions are at room temperature; deviation from the norm will damage the film.

Provided that a few precautions are observed, Polaroid film is capable of good resolution. First, the film should be stored away from heat and humidity—which decrease the film's lifespan. If large amounts of film are purchased, store and desiccate it in a freezer. Before use, let the film stand at room temperature for 24 hr. Boxes which have been opened should be stored in the plastic bag provided by Polaroid—again, to protect against humidity.

An annoying problem is contamination of the camera back rollers by breaking of the developer pod during processing. It is essential that the camara back be removed from the SEM and the rollers routinely cleaned with water. Otherwise, horizontal stripes will be scratched along the emulsion. Do not confuse these with charging artifacts: dirty rollers leave deep scratches, whereas charging is usually a broad band.

Also, Polaroid film contains in the developing pod a caustic jelly which may cause burns if not washed off. Because it is common to contact the developer when separating the film from the packet, always wash your hands after using this film.

PRINTING

The major advantage of using 35-mm or Type 55 Polaroid Film is to obtain a high-resolution negative suitable for enlargement. The processing of these prints is identical to that used in conventional black-and-white photography. Rather than endorse any specific products, the reader should consult a photography textbook, or the following low cost, but informative Kirkpatrick's pamphlet: *Basic Darkroom. Peterson's Guide to Basic Darkroom* (1975), or Eastman Kodak's *Electron Microscopy and Photography* (1973). The former is available at many camera shops, and the latter may be ordered directly from Kodak.

REFERENCES

Adams, A. (1968). *The Negative.* Morgan Press, Hastings-on-Hudson, New York.

Adelstein, P. Z. and D. A. Leister (1963). Nonuniform dimensional changes in topographic aerial films. *Photogram. Engng.* **29**:149.

Altman, J. H. and R. C. Ball (1961). On the spatial stability of photographic plates. *Photo. Sci. Eng.* **5**:278.

Anderson, K. and P. B. Kenway (1967). External photography of the microscope image. *Proc. 25th Ann. EMSA Meet.*

Baggett, M. C. and L. H. Glassman (1974). SEM image processing by analog homomorphic filtering techniques. *IITRI/SEM* p. 199.

Bertaud, W. S., *et al.* (1978) Recording of scanning electron micrographs. In: *Principles and Techniques of Scanning Electron Microscopy* **6**:89. (M. A. Hayat, ed.) Van Nostrand Reinhold, New York.

Boni, A. (1962). *Photographic Literature (1927–1959).* Morgan and Morgan, Dobbs Ferry, New York.

Boyde, A. (1970). Practical problems and methods in the three-dimensional analysis of SEM images. *IITRI/SEM* p. 105.

—— (1971). A review of problems of interpretation of the SEM image with special regard to methods of specimen preparation. *IITRI/SEM* p. 3.

—— (1971). Recording anaglyph stereopairs in the SEM and some other uses of color images in the SEM. *Beitr. Elektronenmikroskop. Kiretabb. Oberfl.* **42**:443.

—— (1974). A stereo-plotting device for SEM micrographs; and a real time 3-D system for SEM. *IITRI/SEM* p. 93.

—— (1974). Photogrammetry of stereopair SEM images using separate measurements from the two images. *IITRI/SEM* p. 101.

—— (1975). Measurement of specimen height difference and beam tilt angle in anaglyph real time stereo TV SEM systems. *IITRI/SEM* p. 189.

—— and P. G. T. Howell (1977). Taking, presenting and treating stereo data from the SEM. *IITRI/SEM* **1**:571.

—— *et al.* (1974). Some practical applications of real time TV speed stereo SEM in hard tissue research. *IITRI/SEM* p. 109.

Carroll, J. S. (1974). *Photographic Lab Handbook.* Amphoto, New York.

Clark–Jones, R. (1958). On the quantum efficiency of photographic emulsions. *Phot. Sci. Engng.* **2**:57.

Eastman–Kodak Co. (1973) *Electron Microscopy and Photography.* Kodak Pub. No. P-236, Eastman Kodak Co., Rochester, New York.

——, (xxxx). *Kodak Films for Cathode-Ray Tube Recording.* Kodak Pub. No. P-37, Eastman Kodak Co., Rochester, New York.

——, (xxxx). *A Kodak Sheet Film for the SEM.* Kodak Pub. No. P-231, Eastman Kodak Co., Rochester, New York.

Engel, C. E., ed. (1968). *Photography for the Scientist.* Academic Press, New York.

Farnell, G. C. and J. B. Chanter (1961). The quantum sensitivity of photographic emulsion grains. *J. Photogr. Sci.* **9**:73.

—— and R. L. Jenkins (1968). The relationship between surface and internal sensitivity of individual grains. *J. Photogr. Sci.* **16**: 256.

—— *et al.* (1973). The computation of the response of model emulsion layers. Part I. Principles. *J. Photogr. Sci.* **21**:118.

Fotland, R. A. (1970). Optically developed free-radical photosensitive materials. *J. Photogr. Sci.* **18**:33.

Hamilton, J. F. and F. Urbach (1966). The mechanism of the formation of the latent image. In: *The Theory of the Photographic Process*, 3rd ed., C. E. K. Mees, and T. H. James, eds. Macmillan, New York.

Hillson, P. J. (1969). *Photography.* Double Day, Garden City, New York.

Howell, P. G. T. (1975). A practical method for the correction of distortions in SEM photogrammetry. *IITRI/SEM* p. 199.

—— (1975). Taking, presenting, and treating stereo data from the SEM. *IITRI/SEM* p. 697.

Hyzer, W. G. (1971). How accurate are photographic measurements? *Res. Dev.* **22**:75.

—— (1975). Sensitivities of photographic materials. *Res. Dev.* **26**:28.

Jacobson, C. I. and R. E. Jacobson (1972). *Developing*, 18th ed. Focal Press, New York.

James, T. H. and G. C. Higgins (1972). *Fundamentals of Photographic Theory*, 3rd ed. Morgan & Morgan, New York.

Kirillou, N. J. (1965). *Problems in Photographic Research.* Focal Press, London.

Kirkpatrick, K. (1975). *Basic Darkroom. Peterson's How-To Photographic Library*, Peterson Publ. Co., Los Angeles.

Langford, M. J. (1972). *Advanced Photography.* Focal Press, New York.

McGee-Russell, S. M. and R. Speck (1970). Color transforms and color fusion stereoscopy for the study of transmission tilt, high voltage and SEM, and freeze-etch replicas. *J. Cell Biol.* **47 (part 2)**:133a.

Mees, C. E. K. and T. H. James, eds. (1966). *The Theory of the Photographic Process*, 3rd ed. Macmillan, New York.

Neblette, C. B., ed. (1976). *Photography: Its Materials and Processes.* Van Nostrand Reinhold Co., New York.

Nemanic, M. (1972). Preparation of red-green lantern slides from SEM micrographs. *Proc. 30th Ann. EMSA Meet.*, p. 412.

—— (1974). Preparation of stereo slides from electron micrograph stereopairs. In: *Principles and Techniques of Scanning Electron Microscopy*, Vol. 1. Hayat, M. A., ed. Van Nostrand Reinhold Co., New York.

Stroke, G. W. and M. Halioua (1971). High-resolution enhancement in scanning electron microscopy by *a posteriori* holographic imaging technique. *IITRI/SEM*, p. 57.

Tovey, N. K. (1971). Soil structure analysis using optical techniques on scanning electron micrographs. *IITRI/SEM*, p. 49.

Wall, E. J. and F. I. Jordan (1974). *Photographic Facts and Formulas*, Amphoto, New York.

3. Introduction to Biological Sample Preparation

SPECIMEN DRYING

The following brief discussion serves simply to familiarize the student with the options available for preparation of a biological sample for SEM. All of these topics will be discussed in depth in later chapters. Regardless of the methodology employed, the desired end-product is a sample dried without artifacts. Consequently, the foremost problem encountered in biological SEM is to bring the sample from its normal hydrated state to the dry state, while still preserving its shape and the distribution of inter- and intracellular contents. Otherwise, a native sample exposed in the SEM to irradiation and high vacuum will be destroyed.

Various mechanisms account for disruption; and although the manifestation of alteration may be similar in different cases, the mechanism may be different. Relatively large cells have a large volume-to-surface ratio, and are most influenced by volume stresses. For example, native bacteria may shrink or collapse, and plant cells (sometimes consisting of 90% water) shrink by plasmolysis when air-dried. As organisms become smaller (*e.g.*, viruses) the volume-to-surface ratio proportionately decreases and surface forces predominate. Analogously, a liter of water requires a rigid container to hold it against gravity, but a drop of water maintains its shape with nothing more than its own surface tension. A similar situation occurs at the ultrastructural level, where the most severe, destructive stress is interfacial tension, defined as the pressure difference between two sides of a liquid meniscus. When cells air-dry, a receding liquid mensicus passes through them, resulting in the build-up of enormous forces. It has been calculated that the stress through an air-drying bacterial flagellum is an incredible 46,000 kg/cm^2.

The easiest way to visualize these forces is as a tiny but forceful hydraulic pump: As the piston—or water-vapor interface—descends, it simply crushes everything in its path. Stress is calculated by having a cylindrical structure of radius r, whose axis is perpendicular to the liquid–vapor interface (Figure 3-1). As the liquid evaporates at the interface, it induces axial compression. This stress on the cross-sectional area of the cylinder, resulting from compression, is calculated from

$$(2\pi \mathrm{rt})\left(\frac{\cos\alpha}{\pi r^2}\right) = \frac{2t\cos\alpha}{r}$$

where

r = radius of cylinder
t = surface tension (dyn/cm)
α = angle of contact of the interface with the cylinder wall
$2\pi r$ = circumference of the cylinder
πr^2 = cross-sectional area of the cylinder

The radius of the cylinder will decrease as the liquid evaporates; *i.e.*, the radius and force approach zero, but the compressional force must first drastically increase. Applying this equation to a microvillus

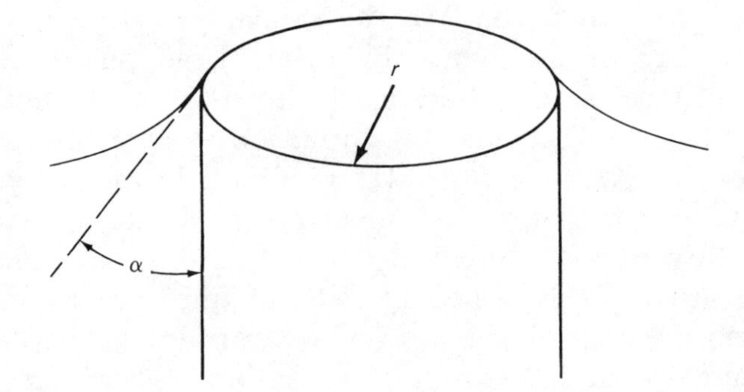

Figure 3-1. The forces of compression caused by air-dying are analogous to the action of a hydraulic pump.

of reasonable proportions ($10^{-5} \times 10^{-4}$ cm; $t_{water} = 70$ dyn/cm; $\cos \alpha = 1$), the stress resulting from the receding water is 2.8×10^7 dyn/cm^2, or 28 atm. All biological specimens, with perhaps the exception of bone and wood, are destroyed under these conditions.

Scheme 3-1 is a flow chart showing the various options available for drying biological samples. They all have in common the goal of reducing or eliminating interfacial tension as it normally exists in biological specimens. Freeze-drying is a physical fixation technique by which tissue is rapidly frozen in liquid nitrogen and, while still frozen, subjected to a low vacuum. The frozen water will then

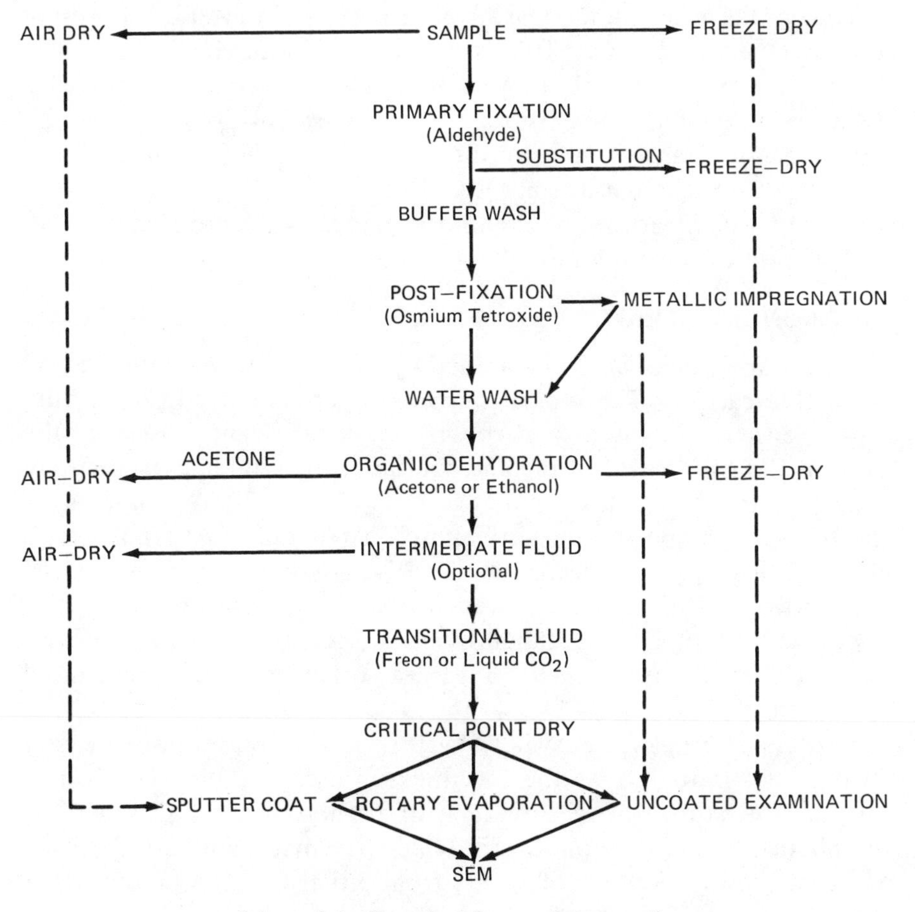

Scheme 3-1. Flow chart for sample preparation.

sublime from the specimen, leaving it dry. This technique does not eliminate interfacial tension, but reduces it.

A second option is to chemically preserve and organically dehydrate the tissue, then dry it by either air-drying or critical point drying. The simplest way to reduce the compression value is to replace the specimen's water with a liquid of lower surface tension. For example, water has a surface tension t of 70 dyn/cm while for acetone it is 23.7 dyn/cm. Therefore, by infiltrating a fixed sample with 100% acetone and air-drying, the compressional forces are reduced by a factor of 3. However, the interfacial tension is still sufficiently high to damage delicate biological specimens.

Today, the only method available to eliminate interfacial tension is critical point drying. This is a technique whereby the liquid in a sample is replaced with another fluid and subjected to a characteristic pressure and temperature. At this critical point, the fluid's liquid phase transforms into the gas phase without the sample's having passed through a meniscus. However, note as in Scheme 3-1, that the tissue must first be chemically fixed. When properly applied, critical point drying is an excellent technique.

SPECIMEN MOUNTING

Before a specimen can be examined by SEM, it must be mounted on a suitable carrier. The typical sample holder is referred to as a stub, and is either aluminum or carbon. To permit manipulation of the stub in the SEM (*e.g.*, tilting and rotating), the sample must securely adhere to the stub. Bulk samples (*e.g.*, tissues) or filtered samples are easily mounted on a stub with double-sided tape. Alternatively, a drop of silver paint (a suspension of silver in amyl acetate) is placed on a stub and the sample placed in the paint. However, the paint tends to creep up the specimen by capillary action and wet it; therefore, it should not be used on a freeze-dried or critical-point dried specimen.

Another alternative, especially useful when x-ray fluorescence is to be studied, is to mount the specimen on a carbon planchet (which sits on the stub) and secure it with carbon paint. This permits simultaneous observation of the electron image and x-ray signal, without the interference of x-rays from either the aluminum stub or silver paint.

Relatively simple techniques are available for applying a suspended particulate sample to the stub. Samples which do not require preservation—that is, which may be air-dried from their native state (*e.g.*, diatoms)—are placed on the stub with a pipette or sprayed on with a nebulizer. Only samples of low concentration may be reasonably studied with this method; otherwise the individual particles clump and obscure detail.

The more preferable method for individual organisms is to filter them on 0.45-μm pore Nucleopore filters. This ensures even distribution across the filter and avoids clumping. Nucleopore membranes are preferred over other types of filters because the specimen lies on the surface, *i.e.*, it does not embed within the filter matrix as it will with paper filters. Thus, higher resolution is possible. More will be said about mounting particulates in Chapter 8.

Whenever filters are employed, they are always mounted on the stub with tape, never with paint. The filter should also lie as flat as possible across the tape without wrinkles. The filters are easily cut with a razor or disks punched out with a cork borer.

COATING TECHNIQUES

Because biological specimens are of such low density, they typically must have their surface density enhanced for sufficient interaction with the electron beam. Otherwise, the sample simply absorbs the beam, resulting in very low resolution. Although uncoated samples are sometimes used for studying insects, more commonly the specimen is coated with a conductive layer. This layer reproduces the surface topography and enhances interaction between the electron beam and the specimen. The three techniques used to increase specimen conductivity are (1) metallic impregnation, (2) sputter coating, and (3) evaporative coating.

Metallic impregnation increases the density of a specimen by reaction with heavy metals. As noted in Scheme 3-1, it commonly follows osmium tetroxide fixation and precedes organic dehydration. Metallic impregnation usually involves ligands which increase the tissue reactivity toward specific heavy metals. Many applications of this technique will undoubtedly evolve—*e.g.*, backscattered electron imaging of reaction sites within tissue. In addition, because the entire specimen, not only the surface, is impregnated, it may be dissected, and different levels examined, without further treatment.

Both sputter coating and evaporative coating involve the deposition of a thin metal film across the specimen's surface. The thin film is sufficiently thick to conduct electrons, but thin enough to ensure high resolution. Sputtering involves etching metal atoms, usually gold or aluminum, from a target, and electromagnetically directing the gold onto the specimen. In comparison, an evaporative coating is formed by heating under high vacuum from a point near the melting point of the metal involved. A thin film then forms on the specimen. Sputtering is an extremely valuable technique applicable to the majority of specimens, whereas evaporation is favored when very minute specimens, *e.g.*, viruses, are being studied. Other pros and cons of these methods are discussed later.

REFERENCES

Albrecht, R. M. and A. P. MacKenzie (1975). Cultured and free-living cells. In: *Principles and Techniques of Scanning Electron Microscopy*, vol. 3, M. A. Hayat, ed. Van Nostrand, New York, p. 109.

——, *et al* (1976). Preparation of cultured cells for SEM: air-drying from organic solvents. *J. Micros.* **180**:21.

Boyde, A. (1972). Biological specimen preparation for the scanning electron microscope: an overview. *IITRI/SEM* p. 257.

—— (1974). Histological and cytological methods for the SEM in biology and medicine. In: *Scanning Electron Microscopy*, O. C. Wells, ed., McGraw Hill, New York, p. 308.

—— (1976). Do's and don'ts in biological specimen preparation for the SEM. *IITRI/SEM* **1**:683.

deHarven, E., *et al.* (1975). New observations on methods of preparing cell suspensions for scanning electron microscopy. *IITRU/SEM* p. 361.

Drier, T. M. and E. L. Thurston (1978). Preparation of aquatic bacteria for enumeration by scanning electron microscopy. *SEM, Inc.* **2**:843.

Echlin, P. and A. J. Saubermann (1977). Preparation of biological specimens for X-ray microanalysis. *IITRI/SEM* **1**:621.

Hayes, T. L. and J. B. Pawley (1975). Very small biological specimens. In: *Principles and Techniques of Scanning Electron Microscopy*, Vol. 3, M. A. Hayat, ed., Van Nostrand, New York, p. 45.

Howard, K. S. and M. T. Postek, (1979). Dehydration of scanning electron microscope specimens—a bibliography. *SEM, Inc.* **2**:892.

Humphreys, W. J. (1975). Drying soft biological tissue for scanning electron microscopy. *IITRI/SEM* p. 707.

Johari, O. and P. DeNee (1972). Handling, mounting, and examination of particles for SEM. *IITRI/SEM* p. 249.

Klainer, A. S. *et al.* (1974). Evaluation and comparison of techniques for examination of bacteria by scanning electron microscopy. *IITRI/SEM* p. 313.

Kormendy, A. C. (1975). Microorganisms. In: *Principles and Techniques of Scanning Electron Microscopy*, Vol. 3, M. A. Hayat, ed., Van Nostrand, New York, p. 82.

Lamb IV, J. C. and P. Ingram (1979). Drying of biological specimens for scanning electron microscopy from ethanol. *SEM, Inc.* **3**:459.

Liepins, A. and E. deHarven (1978). A rapid method of cell drying for scanning electron microscopy. *SEM, Inc.* **2**:37.

Marchant, H. J. (1973). Processing small delicate biological specimens for scanning electron microscopy. *J. Micros.* **97**:369.

Munger, B. L. (1977). The problem of specimen conductivity in electron microscopy. *IITRI/SEM* **1**:481.

Pfefferkorn, G. E. (1970). Specimen preparation techniques. *IITRI/SEM* p. 89.

—— (1973). Techniques for non-conductive samples. *IITRI/SEM* p. 751.

Schneider, G. B. (1976). The effects of preparative procedures for SEM on the size of isolated lymphocytes. *Am. J. Anat.* **146**:93.

Small, E. B. and T. K. Maugel (1978). Observations on the permanence of protozoan preparations for scanning electron microscopy. *SEM, Inc.* **2**:123.

4. Chemical Fixation

A scanning electron microscope (SEM) presents a very hostile environment for biological specimens. Vacuum exposure and electron bombardment reduce the vast majority of untreated (native) specimens to useless waste; consequently, soft biological materials require stabilization and drying prior to SEM examination. The typical preparation sequence for animal or botanical specimens is as follows (Scheme 4-1):

1. *Chemical fixation*: cessation of all cellular activities and preservation of the tissue structure as it was in life.
2. *Chemical dehydration*: substitution of cell water with a reagent miscible with both water and the fluid from which the specimen will be dried.
3. *Drying*: desiccation of the specimen in a controlled manner, while maintaining the same size and shape of the original living material.
4. *Conductivity enhancement*: increasing the electron density (scattering-power) of the specimen surface by sputtered or evaporative coatings; or increasing the scattering power of the bulk specimen by the application of heavy metal salts.

Chemical fixation and dehydration are the topics covered in this chapter. Controlled desiccation by critical-point drying, freeze-drying, or solvent evaporation are covered in later chapters: It is important that the reader note that chemical fixation and dehydration normally precede drying. However, isolated exceptions to this

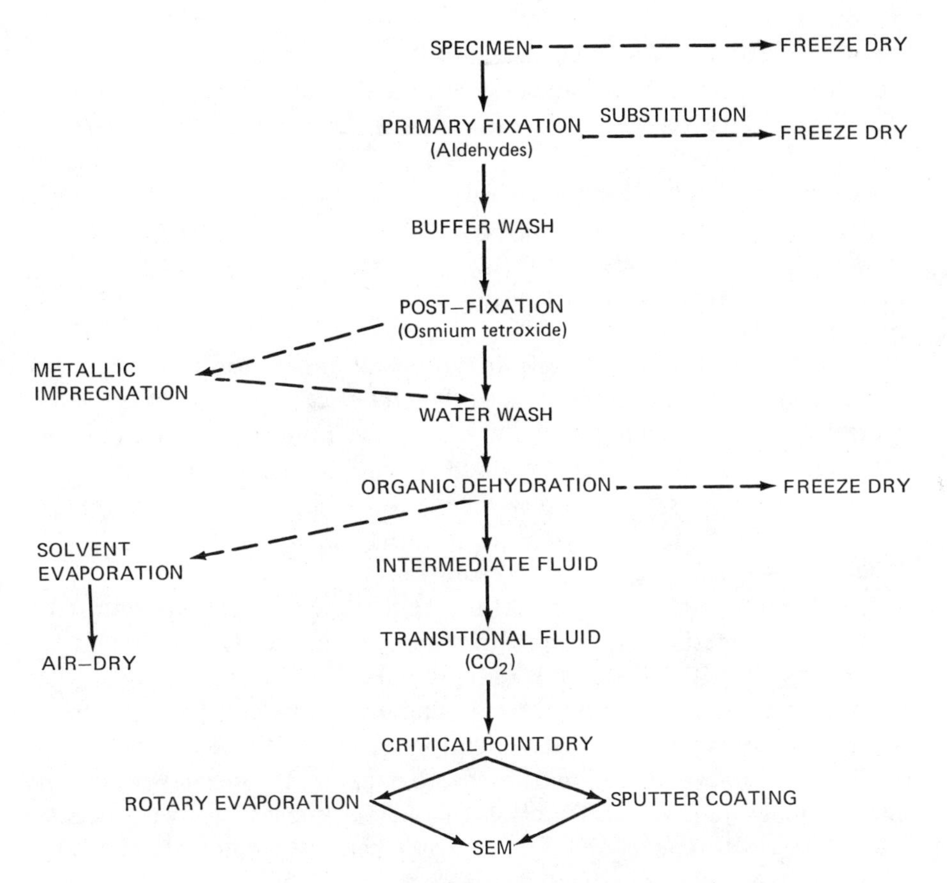

Scheme 4-1. Flow chart for SEM tissue preparation.

rule exist: (1) Organisms possessing a tough exoskeleton (*e.g.*, diatoms) generally do not require fixation (see "Handling Free-Living Cells"), (2) nor does the examination of fresh-frozen botanical specimens (see "Uncoated Specimens"). Finally, because biological specimens possess minimal electron scattering power, conductivity must be enhanced for optimal imaging and resolution. These topics are also considered in future chapters.

CHEMICAL FIXATIVES

Upon reaction with cell macromolecules, organic noncoagulant fixatives preserve ultrastructure by chemically becoming part of the stabilized structure (O'Brien *et al.*, 1973). In so doing, all cell activity immediately stops; cell shape and volume do not change, nor is there displacement or leaching of any cell material (Pentilla *et al.*, 1974 and 1975; Brunk *et al.*, 1975; Nowell and Pawley, 1980). This ideal state has not been achieved; however, microscopists have made and continue to make headway in quantitative understanding of quality preservation. The basic protocol of preservation was founded in light and transmission electron microscopy, primarily because these techniques have been in existence considerably longer than SEM. Therefore, one mode of defining quality preservation is by data correlation with other microscopes (Wetzel *et al.*, 1973; Albrecht and Wetzel, 1979). In general, if the quality of internal morphology as observed by TEM is acceptable according to TEM standards, the surface morphology as observed by SEM should be acceptable. Similarly, light microscope preparations must meet the criteria as laid down for lower levels of resolution.

It is also important to note that the appearance of cells is a function of sex and genetic background (Simson *et al.*, 1978), age (Lawton and Harris, 1978), and the prevailing metabolic state (Gale, 1977).

Another effective method for assessing preservation is to examine a specific tissue type using different preparation methods, all employing SEM as the evaluating instrument. For example, one may measure dimensional changes as a function of buffer type or fixative concentration (*e.g.*, Boyde *et al.*, 1977; Boyde and Macconachie, 1979). Literature searches are, of course, also very useful for corroborating data; for example, one may not have both a critical-point-dryer and freeze-dryer, but perusing of journals very often leads to comparative data.

Chemical fixation normally follows a two-step sequence (double fixation), with primary fixation of proteins and post-fixation of unsaturated lipids by, respectively, an aldehyde (or mixture of aldehydes) and osmium tetroxide. Tertiary fixation with uranyl acetate for stabilization of nucleic acids and glycogen is also popular. It will be seen, however, that this three-phase fixation method is more popular in TEM than in SEM; in the latter situation many researchers are satisfied with aldehyde preservation. Nonetheless, double fixation is recommended to minimize extraction (*e.g.*, Pentilla *et al.*, 1974; Demsey *et al.*, 1978; Mersey and McCully, 1978).

Primary fixation must be initiated before autolytic activities alter the normal living cell (*e.g.*, Wrigglesworth *et al.*, 1970; Schmalbruch, 1980). Thus, the method of introducing the fixative to the tissue is of utmost importance. Basically, a primary fixative may be perfused—*i.e.*, the normal circulatory fluids are substituted with a fixative, resulting in intimate, immediate contact of the fixative with individual cells. Alternatively, the fixative is dripped over the exposed, *in situ* tissue, or finally, the tissue is removed from the organism and immersion fixed. The more rapidly the tissue is penetrated by the fixative, the more rapid is stabilization.

When the fixative contacts the cytoplasm, it will evoke a conversion from the normal semiliquified state to a sponge-like colloidal state (Hayat, 1970). Mersey and McCully (1978) elegantly recorded with Nomarski optics the progress of the fixation front through plant cells, and observed that organelles and inclusions are stabilized only after the front is well beyond their location. After this point, however, organelles are essentially suspended within a three-dimensional network of precipitated proteins. As this real-time event progresses, the pH and tonicity of the fixative must be maintained at physiological values to prevent distortion. Thus, the fixative molecules are carried in a buffer vehicle which accommodates ionic changes during the reaction and prevents shrinkage or swelling of the cells. These factors, their interrelationships, and end-results are the topic of this chapter. The toxicity of fixatives will be considered in their respective discussions; every microscopist should be required to read the SEM safety papers of Humphreys (1977) and Thurston (1978).

Primary Fixation: Aldehydes

The natural ultrastructure of a cell is a function of its protein composition: primary fixation serves to stabilize these macromolecules. Aldehydes very effectively stabilize fine structure by forming both intra- and intermolecular methylene crosslinkages at arginine, cysteine, glutamine/asparagine, histidine, lysine, tryptophan, and tyrosine amino acids (Bowes and Cater, 1968b; Hopwood, 1969a and 1972). The degree of stability of the fixed macromolecules is a function of the specific aldehyde employed, pH, duration of fixation, ambient temperature, and concentration. The commonly used aldehydes in EM are glutaraldehyde, formaldehyde, and acrolein (Table 4-1), or

Table 4-1. Characteristics of Aldehydes Used in SEM.

ALDEHYDE	CHEMICAL STRUCTURE	M.W.	EFFECTIVENESS FOR EM*	SPECIAL CHARACTERISTICS
Formaldehyde	$H_2C{=}O$	30.03	Poor	Rapid penetration
Acrolein	$H_2C{=}CH{-}CHO$	56.06	Good	Rapid penetration and fixation
Glutaraldehyde	$OHC{-}CH_2{-}CH_2{-}CH_2{-}CHO$	100.12	Excellent	Dialdehyde
Glutaraldehyde/ formaldehyde			Excellent	Rapid penetration and fixation of large, bulk tissues
Glutaraldehyde/ acrolein			Excellent	Rapid penetration and fixation of dense tissues

*Overall tissue preservation enhanced by osmium tetroxide postfixation.

a combination of these. Tissue-bound aldehyde groups are referred to as carbonyl compounds and may be detected by the application of Schiff's reagent (Van Duijn, 1961). When observed by light microscopy, Schiff-positive reaction sites are stained pink.

Although aldehydes very effectively stabilize proteins, other macromolecular species are predominantly inert and, consequently, will be extracted either during primary fixation or subsequent treatment (especially dehydration). Extraction is reflected at the SEM level by shrinkage: The degree of overall shrinkage is linearly proportional to the degree of extraction, and Nowell and Pawley (1980) state that the relative amount of a specimen's topographic alteration is a function of the tissue type, homogeneity of the given sample, and overall specimen size. To decrease the gross extraction of unfixed materials, specimens are optimally preserved by post-fixation with osmium tetroxide (*cf.* Schiff and Gennaro, 1979b). The extraction of unfixed material is prevented by monitoring the ionic composition of the buffer (Brunk *et al.*, 1975; Arborgh *et al.*, 1976), the concentration of fixative (Igbal and Weakley, 1974; Thornwaite *et al.*, 1978; Lee *et al.*, 1979), the duration of fixation (Hopwood, 1970b and 1972), and by avoiding overdehydration of the tissue (Cohen, 1979). In practice, all of these interrelated factors affect preservation; these correlations will be presented below.

Buffers

Fixative molecules are carried to cells in a buffer vehicle. Although all cells have a natural buffering system, it is overwhelmed by fixation activities. As mentioned above, the application of an aqueous (non-buffered) primary fixative radically changes the physiological pH of a cell, and osmotic effects will distort gross and ultrastructural morphology by shrinkage or swelling (*e.g.*, Palade, 1952; Caulfield, 1957; Tooze, 1964; O'Brien *et al.*, 1973; Arborgh *et al.*, 1976). Claude (1961) and Malhotra (1962) determined that pH dropped from 6.2 down to 4.4 when cells were fixed in unbuffered osmium. Any drop in pH, whether it be by the above mechanism or during normal cell death, reduces macroproteins into small, ionized fragments: Ultrastructure is a function of proteins, and if protein conformation is lost, disruption at the morphological level will clearly be evident

(Wrigglesworth *et al.*, 1970). This effect is similar to natural cell death, where a drop of cell pH by macromolecular dissociation initiates autolysis. Therefore, artificial buffers serve to accommodate the ions generated at the reaction front at or near physiological values. Tonicity (the volumetric response of cells immersed in a solution–*i.e.*, swelling, shrinkage, or no effect) is also monitored by the buffer. Either the buffer composition is such that it serves this purpose, or additional electrolytes (*e.g.*, magnesium or calcium salts) or nonelectrolytes (glucose or sucrose) are added to the vehicle.

The pH and ionic composition of fixative buffers are maintained as close as possible to physiological values for a given specimen type. Generally, plant tissues fall within pH 6.8–7.1, and animal tissues within pH 7.0–7.4: The literature should be checked for specific pH, although "physiological range pH" is acceptable (Glauert, 1975; Schiff and Gennaro, 1979a and b). Tonicity, however, is not as clearly defined, yet a number of factors have been experimentally demonstrated (Shaw, 1960). First, fixatives may alter, but not necessarily destroy, the selective permeability of cell membranes (Bone and Denton, 1971; Carstensen *et al.*, 1971; Hopwood, 1972; Schiff and Gennaro, 1979a and b); thus, the tonicity of the primary fixative is critical, but that of the postfixative is less critical (Fahimi and Drochmans, 1965b, Jard *et al.*, 1966; Moss, 1966; Jost *et al.*, 1973; Boyde, 1976; Boyde and Macconachie, 1979; Schiff and Gennaro, 1979a; Tobin, 1980). Second, some researchers maintain that only the osmolarity of the buffer without the addition of aldehyde is osmotically important (Bone and Denton, 1971; Bone and Ryan, 1972; Arborgh *et al.*, 1976; Boyde, 1976; Boyde and Macconachie, 1979). In comparison, another faction of researchers contend that the aldehyde plus the buffer are osmotically active (Boyde and Vesely, 1972), particularly when the buffered aldehyde is applied directly to the surface of interest (Matthieu *et al.*, 1978). In this situation (*i.e.*, immersion fixation) the buffered aldehyde should be isotonic (Hayat, 1978): When the area of interest is beneath the surface and will be exposed by fracturing after fixation (Boyde, 1975; Flood, 1975), hypertonic buffered fixatives may be used because they will become isotonic by dilution with cell fluids (Karnovsky, 1965; Schneeberger-Keeley and Karnovsky, 1968; Davey, 1973).

Finally, numerous researchers have demonstrated that tonicity of the buffered fixative is related to the concentration of fixative molecules (Fahimi and Drochmans, 1965a,b; Maunsbach, 1966; Maser *et al.*, 1967; Bone and Denton, 1971; Bone and Ryan, 1972; Igbal and Weakley, 1974; Pilstrom and Nordlund, 1975; Pexieder, 1976; Thornwaite *et al.*, 1978; Lee *et al.*, 1979; Schiff and Gennaro, 1979b). These authors have empirically demonstrated that high concentrations of aldehydes induce shrinkage–*i.e.*, the tissue proper is used as a natural osmometer (Mathieu *et al.*, 1978). Thornwaite *et al.* (1978) and Lee *et al.* (1979) took a novel approach to determining the optimum fixative concentration by monitoring high-resolution electronic cell volume spectra.

The tonicity of a buffer is measured by freezing-point depression (Tahmisian, 1964) or with an osmometer (Maser *et al.*, 1967). Rarely is it necessary for the microscopist to work with such exact methods: The formulations of common buffers and their tonicity are presented in the Appendix. If the buffer alone does not achieve the desired tonicity, it may be adjusted by adding nonelectrolytes (*e.g.*, sucrose; Caulfield, 1957) or electrolytes (*e.g.*, $CaCl_2$; Epling and Sjostrand, 1962). Monovalent cations will avoid excess protein precipitation (Millonig and Marinozzi, 1968), while divalent cations, such as magnesium and calcium, may irreversibly shrink the tissue (Boyde and Macconachie, 1979; McKinley and Usherwood, 1978). It is known that the buffer actively participates in the fixation process (Trump and Ericsson, 1965; Schiff *et al.*, 1976; Gennaro *et al.*, 1978; Schiff and Gennaro, 1979a,b), and that the ionic composition of the buffer affects preservation (Maunsbach, 1966; Schiff *et al.*, 1976; Kuran and Olzewski, 1977; Gale, 1977). Some of these problems may be minimized by incorporating more inert nonelectrolytes in the buffer (Wood and Luft, 1963; Millonig, 1964). Preferably, however, the buffer formulation itself should function as a moderator of both pH and osmolarity (Tobin, 1980).

Schiff and Gennaro (1979b) provide an excellent review of the various buffers used in SEM; general discussions of biological buffers are available in Good *et al.*, (1966) and Dawson *et al.*, (1969). Only a few buffers are commonly employed, and have the following characteristics: stability and effectiveness at physiological pH (6.0–8.0); ionic composition similar to natural buffers; and, in some cases, nontoxicity. Whereas aquatic organisms may be fixed with a vehicle consisting of their natural environment (seawater; Maugel *et al.*, 1980), most other specimen types are buffered with phosphates (Millonig, 1961 and 1964), sodium cacodylate (Sabatini *et al.*, 1963), *sym*-collidine (Gomori, 1946), or PIPES [Piperazine-N-N′-bis(2-ethanol sulfonic acid)] (Salema and Brandao, 1973; Rudman *et al.*, 1978). Historically, other buffers have been used, but are currently rejected because they exert undesirable effects on fixation: Examples are sodium bicarbonate, which destroys microtubules (Schultz and Case, 1968); and veronal acetate, which exerts an undesirable barbituate effect (Palade, 1952). A number of authors have systematically evaluated different buffers for given tissues (Wood and Luft, 1963, 1965; Gil and Weibel, 1968, 1969–1970; Gale, 1977; Kuran and Olzewski, 1977), and these effects will be discussed with specific fixatives.

Phosphate buffers are popular because they are similar to natural body fluids; they contain both mono- and divalent cations, and therefore osmolarity may be readily adjusted, and, if desired, readjusted by the addition of nonelectrolytes. A number of sodium phosphate formulations are included in the "Appendix." The choice of one is a function of the desired tonicity (Busson–Mabillot, 1971). These buffers are also inexpensive, nontoxic, and compatible with both primary and postfixatives. On the other hand, phosphate buffers support microorganism growth and may embrittle the specimen. In some situations, a precipitant was observed within cells (Gil and Weibel, 1968; Kuthy and Csapo, 1976).

Sodium cacodylate buffers yield results similar to phosphates, although tonicity is adjusted by adding sucrose, glucose, or sodium chloride (Brunk *et al.*, 1975). Because this buffer contains arsenic, it must be handled in a fume hood and the user should wear gloves (Weakley, 1977). Furthermore, Schiff and Gennaro (1979b) caution that toxicity of sodium cacodylate relative to a cell should not be ignored. This same effect accounts for this buffer's ability to remain uncontaminated over time, although age is accompanied by a drop in pH (Glauert, 1975).

Sym-Collidine (2,4,6-trimethyl pyridine; Gomori, 1946; and Bennett and Luft, 1959) is an alternative to phosphate buffers. While quality preservation of specimens is obtained, this buffer must be very pure to be effective. However, it is toxic, quite expensive, and does not exhibit any major advantages over phosphate or cacodylate buffers.

PIPES buffers enjoyed short-lived popularity which may be rekindled in the future; governmental agencies are investigating the carcinogenic effects of PIPES. Schiff and Gennaro (1979b) demonstrated that this buffer is an excellent substitute for the above buffers, and found that primary fixation was enhanced to such a degree that postfixation was not necessary for good preservation. According to Schiff and Gennaro (1979a), less lipid was extracted, extensive lengths of membranes were continuous, and membrane blisters were not observed (although an artifact is sometimes observed in doubly fixed tissues–Shelton and Mowczko, 1978).

The effects of buffer/fixative combinations will be considered in the discussion of specific fixatives. The reader should consult the following reviews for specific information on buffer type and tonicity: animal tissues, Nowell and Pawley (1980); bacteria, Watson *et al.*, (1980); botanical specimens, Falk (1980); and cultured cells, Shay and Walker (1980).

Glutaraldehyde

The aldehyde most effective for stabilizing proteins and, thus, ultrastructure, is glutaraldehyde (glutaric acid dialdehyde, $C_5H_8O_2$; Sabatini *et al.*, 1962 and 1963). In comparison to other commonly used primary fixatives, it is unique in being a dialdehyde–*i.e.*, each molecule contains two reactive groups, either or both capable of participating in fixation (Cater, 1963 and 1965; Bowes and Cater, 1966 and 1968; Wold, 1967; Richards and Knowles, 1968; Jansen *et al.*, 1971).

At the submacromolecular level, glutaraldehyde forms intra- and intermolecular methylene crosslinks between reactive side chains of proteins: Severe conformational changes do not occur because peptide bonds are unaffected (Habeeb and Hiramoto, 1968). This has been confirmed by high-resolution x-ray diffraction of reacted protein crystals (Quiocho *et al.*, 1967; Moretz *et al.*, 1969b). Fixed proteins resist extraction, but unfixed macromolecules may be extracted (Richards and Knowles, 1968; West and Mangan, 1970).

Glutaraldehyde crosslinks reactive amino acids (Hopwood *et al.*, 1970; Alexa *et al.*, 1971; Chisalita *et al.*, 1971; Hopwood, 1970a,b, and 1972; Conger *et al.*, 1978) in a two-step sequence:

1. Aldol condensation
   a. Two glutaraldehyde molecules polymerize by an aldol condensation (Aso and Aito, 1962; Bowes and Cater, 1966; Richards and Knowles, 1968).

$$2(OHC{-}C_3H_6{-}CHO) \xrightarrow{OH^-} OHC{-}C_3H_6{-}CH{=}C^2H_4{-}CHO$$

   b. Additional reaction between the above product and other glutaraldehyde molecules yields an aldol.

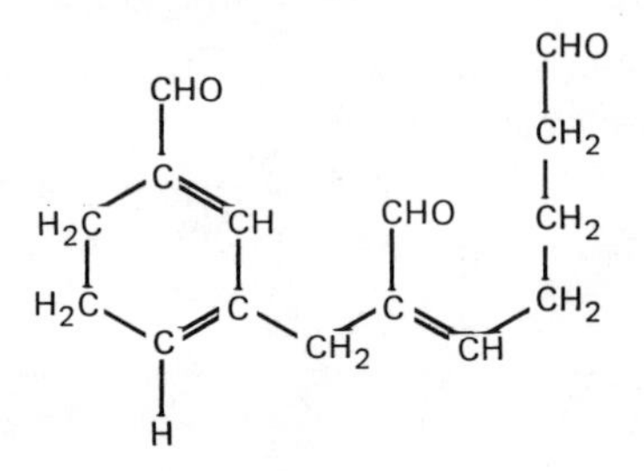

2. The aldol reacts with nitrogen atoms of amino groups during fixation (Richards and Knowles, 1968).

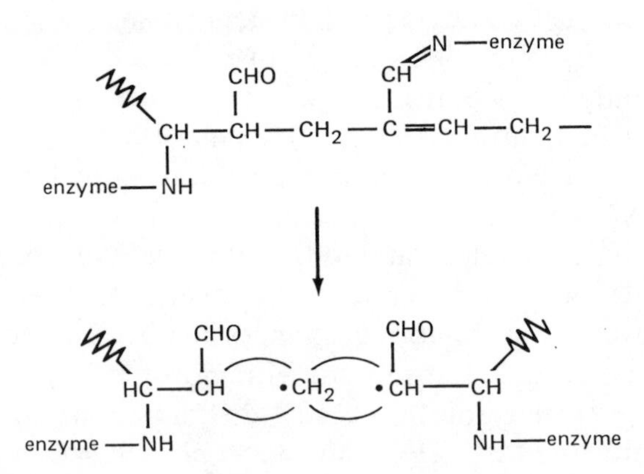

Other potential reactive sites are N-terminals of some peptides, active hydrogen groups, and imino groups (Habeeb and Hiramoto, 1968).

Jansen *et al.* (1971) and Tomimatsu *et al.* (1971) confirmed that stabilization, indeed, involves crosslinking and polymerization; which reaction occurs first probably is a function of the degree of polymerization (which increases with age) of the glutaraldehyde in solution. Robertson and Schultz (1970) indicate that stabilization is a reaction with polymeric glutaraldehyde; and Perrachia and Mittler (1972) introduced a method of fixation that applies this concept. Their method involves initial fixation with neutral buffered glutaraldehyde where the monomeric species is predominant; and because the molecular weight of the monomer is relatively low, the fixative molecules will rapidly penetrate the cells. Shortly afterwards, this solution is replaced with fixative of higher pH, thus inducing polymerization of the glutaraldehyde. This in turn promotes crosslinking and increases stability. Similarly, Hopwood (1970a,b and 1972) noted that the degree of crosslinking increases dramatically between the first and fourth hour of fixation—again indicating that polymerization precedes crosslinking.

Other classes of macromolecules may react with glutaraldehyde, but none do so as thoroughly as proteins. Both unsaturated lipids and saturated lipids are predominantly inert toward glutaraldehyde (Levy, 1965); these will be fixed during exposure to osmium tetroxide. Selected, reactive groups on nucleic acids lend some stability (Millonig and Marinozzi, 1968; Hopwood, 1972 and 1975), but much better preservation occurs during tertiary fixation with uranyl acetate (Huxley and Zubay, 1961; Hayat, 1968; Sechaud and Kellenberger, 1972). Phospholipids are preserved by both osmium tetroxide and uranyl acetate (Silva *et al.*, 1971). Jones (1972) determined that some reactivity exists between fatty acids and glutaraldehyde. Carbohydrates, especially glycogen, were demonstrated to be only partially extracted after glutaraldehyde fixation (65% retained with glutaraldehyde *vs.* 75% retention with formaldehyde; Hopwood, 1967). Apparently, glycogen is somewhat modified, but its native insolubility also prevents excessive extraction. De Bruijn (1973), Czarnecki (1971), and Robertson *et al.* (1975) noted that glycogen is modified by glutaraldehyde, osmium tetroxide, and uranyl acetate.

The effects of glutaraldehyde fixation on both internal and external ultrastructure have been well defined for many tissues; most of the following discussion should be taken in the context that postosmication is usually included for optimal preservation. The general effects of glutaraldehyde on botanical internal ultrastructure are discussed by Ledbetter and Gunning (1963), Moss (1966), Feder and O'Brien (1968), West and Mangan (1970), O'Brien *et al.* (1973), and Mersey and McCully (1978): The evaluation of external plant morphology after glutaraldehyde fixation is summarized by Falk *et al.*, (1971), Parsons *et al.*, (1974), Lawton and Harris (1978), and Falk (1980). TEM of fixed animal tissues is appraised by Baker (1965), Ericsson *et al.* (1965), and Busson-Mabillot (1971); and observations with the SEM are summarized by Cohen and Shaykr (1974), Thornwaite *et al.* (1978), and Nowell and Pawley (1980). Tissue and cell cultures are documented by Boyde and Vesely (1972), Buckley (1973a,b), Johnson (1978), Lee *et al.* (1979), Collins *et al.* (1980), and Shay and Walker (1980). Effects on bacteria may be found in Glauert and Thornley (1966) for TEM, while SEM morphology is summarized by Williams *et al.* (1973), Kormendy (1975), Drier and Thurston (1978), and Watson *et al.* (1980). Glutaraldehyde fixation and its effects on embryonic tissues are discussed by Kalt and Tandler (1971), Boyde and Macconachie (1979), and Waterman (1980). Finally, because subcellular fractions are being investigated with the SEM, the effect of glutaraldehyde on preservation of organelles is important. General membrane structure has been studied by Pladellorens and Subirana (1975) and van Deurs and Luft (1979); nuclei examined by Skaer and Whytock (1977) and Willison and Rajaraman (1977); and mitochondria by Stoner and Sirak (1969) and Pilstrom and Nordlund (1975)—who also evaluated rough endoplasmic reticulum.

The following list serves to familiarize the reader with specific tissues that have been evaluated following glutaraldehyde fixation. By no means is the list complete, but it may assist those individuals who are unfamiliar with either classical or current literature. Many tissues are discussed in the 1978–1980 volumes of *SEM, Inc.*, which are the proceedings from an international symposium. This literature is an excellent compilation of state-of-the-art methods.

TISSUE	REFERENCES
Blood cells	Hirsch and Fedorko (1968), Hirsch *et al.* (1968), Morel *et al.* (1971): also see Chapter 8.
Blood vessels	Buss *et al.* (1976), Clark and Glagov (1976)
Brain	Schultz and Karlsson (1965), Johnston and Roots

	(1967), Bodian (1970), Deutsch and Hillman (1977), Stumpf *et al.* (1977)
Kidney	Trump and Ericsson (1965), Maunsbach (1966), Johnston *et al.* (1973), Larsson (1975)
Liver	Ericsson and Biberfeld (1967), Pilstrom and Nordlund (1975)
Lung	Kuhn and Finke (1972), Mathieu *et al.* (1978)
Muscle	Forbes and Sperelakis (1971), Schmalbruch (1980)
Ovary	Igbal and Weakley (1974)
Testis	Nowell and Faulkin (1974)

For the data presented above, we assume that optimal conditions were in force: The practical effects which control the quality of glutaraldehyde fixation are ambient temperature, concentration, and duration of exposure; as well as buffer osmolarity and pH. For a given tissue, literature searches often reveal acceptable conditions, but experimental evaluation remains the norm. Nonetheless, several generalizations may be made.

Primary fixation is normally carried out at 0–4°C to minimize autolysis (autolytic enzymes are most reactive at warm temperatures), ensure uniform fixation (particularly when osmication will be conducted), and reduce extraction from overfixation (Nowell and Pawley, 1980). Cooling of the tissue is always recommended after the tissue has been removed from the bulk organism, but perfusion is normally conducted at deep body temperature (Boyde, 1976). Mersey and McCulley (1978) did not observe any difference between warm and cold fixation of plant materials, provided that fixation was conducted for 2 hr. Similarly, Watson *et al.* (1980) and Maugel *et al.* (1980) note that fixation at room temperature is acceptable. On the other hand, this author recommends that the specimen be fixed at its normal temperature for $\sim$15 min, then gradually cooled to 0–4°C: It is known that cold fixation is optimal for TEM (Chambers *et al.* 1968; McDowell and Trump, 1976). The specimen is then rewarmed to room temperature during the final chemical dehydration step. A striking example in regard to avoiding thermal shock was shown by Langenberg (1979), who demonstrated that fragile viral inclusions in plants were preserved only when the tissue was prechilled and fixed in the cold.

An adverse effect of cold fixation is that both the penetration and fixation rates proportionately decrease with temperature (Flitney, 1966; Ericsson and Biberfeld, 1967; Hopwood, 1969a, 1970b, 1972, *et al.*, 1970; Falk *et al.*, 1967). Hopwood (1967a) measured the penetration depth of 4% glutaraldehyde into liver, and found that the fixation depth was 4.5 mm and 2.5 mm at, respectively, room temperature and 0–4°C. Subsequently, he has also demonstrated that the degree of crosslinking increased dramatically between the first and fourth hours of fixation (Hopwood, 1970b and 1972). In practical terms, this implies that fixation is uniform for relatively thin slices ($\sim$1 × 3 × 5 mm) when conducted at 0–4°C with moderate concentrations (2–4%) (Fahimi and Drochmans, 1965a,b; Schultz and Karlsson, 1965; Pentilla *et al.*, 1974 and 1975). Concentrated glutaraldehyde ($\sim$37%) severely shrinks tissues, while dilute solutions induce swelling (Fahimi and Drochmans, 1965a) and induce erratic fixation (Igbal and Weakley, 1974). Furthermore, lipid retention is highest at moderate concentration (Busson–Mabillot, 1971; Matthieu *et al.*, 1978).

The effects of tonicity were discussed earlier and will not be repeated here. Note that glutaraldehyde may be prepared with any common buffer; that the final fixative should be isotonic when specimens are fixed by immersion and when the surface of interest is exposed during fixation; and, finally, that slightly hypertonic fixatives may be used for perfusion or when the interior of the tissue is exposed after fixation for SEM imaging. Always maintain pH at physiological value for the given tissue.

Aqueous glutaraldehyde is commercially available either in bulk volumes (25–70%; pints) or in inert atmosphere ampoules of lower concentration (8, 25, 50, 75%; 10 ml/vial). Large volumes of stock glutaraldehyde are prone to undesired polymerization which will degrade preservation (Aso and Aito, 1962; Fahimi and Drochmans, 1965a; Hopwood, 1967b; Jones, 1974; Weakley, 1974; Gillett *et al.*, 1975): The smaller volumes held in an inert atmosphere are much purer. The stock and buffered solutions should be stored in the dark and refrigerated to minimize polymerization. A drop in pH of the stock solution enhances polymerization; in that case, the solution should be discarded. Always wear gloves and work in a fume hood to avoid allergic reactions, which may develop with chronic exposure to any aldehyde. If tannin-mediated metallic impregnation is desired, the fixative is modified at this point. Refer to Chapter 10 for the details of this method.

Formaldehyde

The most common noncoagulant fixative encountered in biology is formaldehyde, CH_2O. It is a monoaldehyde of very low molecular weight (30.03) and, thus, is capable of rapidly penetrating tissues (Baker and McCrae, 1966; Deutsch and Hillman, 1977). Aqueous solutions of formaldehyde (naturally a gas) are referred to as formalin; the solid polymer is paraformaldehyde, $(CH_2O)_n$. Wolman (1955), Sjostrand (1956), and Holt and Hicks (1961) noted that formalin-fixed tissues do not exhibit optimal preservation as resolvable by the electron microscope; numerous other researchers have confirmed this early data at the ultrastructural level (Sabatini *et al.*, 1963; Baker, 1965; Ericsson *et al.*, 1965; Ericsson and Biberfeld, 1967; Gonzalez–Aguilar, 1969; Moretz *et al.*, 1969b; Pladellorens and Subirana, 1975; Deutsch and Hillman, 1977; Mersey and McCully, 1978). Glauert (1975) indicates that methanol-free formalin is much better for fixation rather than concentrated solutions; unless purification of the formalin is conducted, poor preservation results. Her purification method is included in the Appendix. However, a mixture of glutaraldehyde and formalin is very useful; mixed aldehydes are considered under "Comparison of Aldehydes." Also, tissues stored in buffered formalin show no additional artifacts than those carried through the entire processing procedure (Yamamoto and Rosario, 1967; Artvinli, 1975).

Because pure formalin is only used in rare situations, its chemistry will not be dwelt upon: When a mixture of aldehydes is used for primary fixation, the glutaraldehyde fraction predominates in cross-linking. The chemistry of pure formalin reactions may be found in Ealker (1964), Flitney (1966), Ericsson and Biberfeld (1967), Hopwood (1970b), and McDowell and Trump (1976).

The reactions between proteins, especially collagen, and formalin have been thoroughly investigated by the tanning industry (Bowes and Kenten, 1949; Cater, 1963; Bowes *et al.*, 1965; Bowes and Cater, 1968). As with glutaraldehyde, proteins are stabilized by formalin in a two-step process: First, CH_2O molecules react with free amino groups, resulting in the formation of a methylol group; second, the methylol groups condense with phenols, imidazoles, and indoles via crosslinking methylene bridges (Lojda, 1965). Some reaction between formaldehyde and other macromolecules occur, but these reactions are reversible by hydrolysis [fatty acids–Jones and Gresham (1966), Jones (1969 and 1972); polynucleotides–Haselkorn and Doty (1961); nucleosides–Eyring and Ofengand (1967); carbohydrates–Hopwood (1967a); lipids–Wolman (1955), Wolman and Greco (1952), Jones and Gresham (1966)].

Acrolein

Another useful monoaldehyde is the powerful oxidizer acrolein (acrylic aldehyde, C_3H_4O), introduced by Luft (1959) for light and electron microscopy. Its potency is a function of both its molecular structure and low molecular weight (Table 4-1); acrolein's penetration and fixation rate are significantly higher than any other aldehyde (Luft, 1959 and 1972; Flitney, 1966). In practical terms this means that dilute fixative and short-duration exposure are necessary; an increase of either induces rapid extraction of unfixed cell material (Sabatini *et al.*, 1963 and 1964; Saito and Keino, 1976). In fact, Landis *et al.* (1980) obtained good preservation of bone when it was fixed by acrolein vapors alone. As with formalin, a mixture of acrolein and glutaraldehyde is preferred over pure acrolein. Mersey and McCully (1978) note that this mixture is very effective for rapidly penetrating and stabilizing plant materials possessing dense walls.

Proteins are stabilized by reaction between acrolein and sulfhydryl, aliphatic NH_2 and NH, and imidazole groups (van Duijn, 1961). Unsaturated fatty acid stabilization occurs (Jones, 1969), and acrolein is soluble in lipids (Norton *et al.*, 1962; Schultz and Case, 1968). However, overoxidation by acrolein will result in severe extraction of both fixed and unfixed macromolecules (Sabatini *et al.*, 1963). The user must also be very careful when handling acrolein; Thurston (1978) indicates that the threshold limit value for acrolein is 0.1 ppm (as compared to 2 ppm for glutaraldehyde or formaldehyde). Consequently, acrolein is only handled in a fume hood with the user wearing rubber gloves (Albin, 1962).

Acrolein is commerically available in vials and under an inert atmosphere; working buffered solutions should be prepared just prior to use in order to avoid storage problems and ensure contaminant-free fixative (Albin, 1972). Contamination is indicated by cloudiness or a drop in pH (Hayat, 1970). Waste acrolein is neutralized by combination with 70% sodium bisulfite.

Comparison of Aldehydes

To a large degree, the quality of preservation is a function of an aldehydes's rate of penetration, fixation, crosslinking, and bond stability within the specimen (Table 4-2). The penetration rate is very important, because all the other parameters presuppose that the aldehyde has infiltrated the bulk tissue and entered individual cells. Therefore, a major factor influencing penetration rate is the method of fixation. When the fixative is introduced to the tissue via the organ's vasculature (perfusion fixation), intimate, nearly immediate contact between fixative molecules and cells is established: The penetration rate is enhanced by perfusion fixation. In the opposite direction, *i.e.*, introducing the tissue to the fixative by immersing the specimen, the fixative, under its own power, must penetrate successive layers of cells while simultaneously entering individual cells. Thus, either method must be considered as a real-time event, with penetration significantly slower during immersion fixation. This is a major factor over which the researcher has control.

Other factors influencing penetration rate, and which also are within user control, are the ambient temperature and buffer tonicity.

Table 4-2. Comparison of Aldehydes.

	RELATIVE RATES			
ALDEHYDE	PENETRATION	FIXATION	CROSSLINKING	STABILITY
Glutaraldehyde	Slow	Moderate	Excellent	Excellent
Formaldehyde	Rapid	Good	Poor	Poor
Acrolein	Very rapid	Very Rapid	Moderate	Moderate
Glutaraldehyde/ formaldehyde	Moderate	Good	Excellent	Excellent
Glutaraldehyde/ acrolein	Rapid	Rapid	Excellent	Excellent

Penetration decreases with temperature and with isotonic buffers, but the reasons why these are controlled were indicated earlier. To avoid damage, primary fixation should be conducted at 0–4°C with isotonic fixatives (*cf.* Zeikus and Aldrich, 1975).

The penetration rates of aldehydes are also a function of molecular weight: Penetration rate is inversely proportional to molecular weight. Thus, formalin is very rapid, second is acrolein, and glutaraldehyde is least rapid (Feder and O'Brien, 1968). The rate of penetration of acrolein is closely related to its oxidative capacity–*i.e.*, fixation occurs so rapidly that it may set up a difference in osmotic pressure at the fixation front, and the cells themselves become readily permeable. Fixation rates thus follow this sequence from rapid to slow: acrolein, formaldehyde, and glutaraldehyde (Flitney, 1966; Hopwood *et al.*, 1970). However, the stability of reaction (*i.e.*, ability to resist displacement during subsequent treatment) is led by glutaraldehyde, acrolein, and formaldehyde (Carstensen *et al.*, 1971). Bowes and Cater (1966 and 1968) demonstrated that independently of pH, significantly more than formaldehyde, glutaraldehyde was bound to collagen and introduced more crosslinkages (also see Trnavska *et al.*, 1966; and Alexa *et al.*, 1971).

Given pure aldehyde fixatives, glutaraldehyde is most effective for electron microscopy, especially when followed by postosmication (Sabatini *et al.*, 1963). However, it should be clear that a mixture of aldehydes offset the disadvantages of any one fixative. Karnovsky (1965) formulated a mixed glutaraldehyde/formaldehyde fixative that is used in pathology (McDowell and Trump, 1966), microbiology (DeBault, 1973), general preparation of animal tissues (Schneeberger-Keeley and Karnovsky, 1968; Nowell and Pawley, 1980), and as a storage solution (Tyler *et al.*, 1973). Karnovsky's fixative may be used to fix large blocks of tissue homogeneously, because the formaldehyde molecules rapidly penetrate and fix cells–thus making it easier for the large glutaraldehyde molecules to enter cells: Because the latter affords more stable bonds, it will displace bound formaldehyde and yield excellent stabilization.

A similar pathway is followed when mixtures of acrolein and glutaraldehyde are used for fixation. Kalt and Tandler (1971) used this mixture for embryo preservation; and Mersey and McCully (1978) and Falk (1980) highly recommend this mixture for preserving botanical specimens possessing a dense wall. Although no quantitative descriptions of cell shrinkage have been published regarding glutaraldehyde/acrolein fixation, the user should recall the effects of extraction by acrolein. Also, all precautions noted earlier must be enforced.

Methods of Primary Fixation

A number of routes are available for introducing a fixative to a tissue. Vascular perfusion involves replacement of the tissue's normal circulatory fluids with the fixative; a modification is microinjection of the fixative into the organ or small organism (*e.g.*, embryo) of interest. Another method involving sacrifice of the organism is drip fixation *in vivo*, where the desired tissue is exposed and the fixative is slowly poured over the tissue; subsequently the tissue is dissected out and fixation is continued by immersion. Tissues are fixed by immersion simply by placing a specimen (*e.g.*, biopsy) into a vial holding fixative. These methods apply to bulk plant and animal specimens; refer to Chapter 8 for other methods (*e.g.*, bacteria and other unicellular organisms).

Regardless of which method is used for fixation, it should be initiated rapidly to avoid autolytic effects (Schmalbruch, 1980). Mechanical damage must be avoided, particularly when the exposed surface is the surface of interest (Boyde, 1976). The effects of various anaesthetics should be considered (Altman and Dittmer, 1973; Barnes and Etherington, 1975); sodium pentobarbital as a general anaesthetic is commonly used. Spinal dislocation or beheading of the animal may be used, provided that the desired tissue is rapidly exposed and fixed.

Experimental animal tissue may be fixed by a combination of drip and immersion fixation (Nowell and Pawley, 1980). When the area of interest is close to the organ surface, chilled fixative is dripped over the organ: The fixative will penetrate the tissue and mechanically stabilize it, protecting against damage during removal (Maunsbach, 1966). The fixative should be isotonic when exposed surfaces will be examined; hypertonic solutions, which increase the penetration rate, may be used if the tissue will be subsequently fractured and a new surface exposed for examination (Nowell and Pawley, 1980). After approximately 15 min, the tissue is removed and immersion fixation conducted for 1.5–2 hr at 0–4°C. Tissues possessing a mucus layer (*e.g.*, small intestine) should be gently irrigated with saline or buffer prior to fixation, or during fixation (Eisenstat *et al.*, 1976); if mucus persists it may be teased off during dehydration. Fixation with buffered glutaraldehyde (∿3%) or glutaraldehyde/formaldehyde (4:1, v/v) is recommended. When it is neither practical nor convenient to sacrifice the animal, the tissue is fixed by immersion in cold fixative for ∿2 hr. The tonicity factors noted above also apply in this situation.

Plant tissues are also amenable to combined drip/immersion fixation, again assuming that the surface of interest is close to the external topography. If drip fixation is inconvenient, and the surface of interest is naturally exposed (*e.g.*, a leaf), a region of the tissue may be isolated by a continuous ring of petroleum jelly, the area filled with fixative, and then removed and fixed by immersion. According to Falk (1980) and Mersey and McCully (1978), fixation with glutaraldehyde (∿3%) or a mixture of glutaraldehyde/acrolein for ∿2 hr at room temperature, are standard methods. Because plant cells naturally possess a high internal osmotic pressure, fixative tonicity should be carefully monitored.

The above techniques generally provide adequate preservation, but a number of situations arise where vascular perfusion is preferred (Gertz *et al.*, 1975). First, central nervous system tissues rapidly succumb to anoxia and must be fixed prior to the initiation of damage (Palay *et al.*, 1962; Karlsson and Schultz, 1965; Schultz and Karlsson, 1965; Kalimo, 1976). Second, when the area of interest is deep within an organ (*e.g.*, kidney medulla) optimal preservation is only achieved by perfusion (Maunsbach, 1966; Johnston *et al.*, 1973; Larsson, 1975). Albrecht and Wetzel (1979) carried this example further by stating that any solid tissue is optimally preserved by perfusion.

Perfusion apparatuses have been described by Palay *et al.* (1962) and Gil and Weibel (1969–1970). A simple apparatus employing one gravity-fed reservoir usually is sufficient. It is critical, however, that the perfusion be gentle rather than forceful—which could damage ultrastructure (Van Harreveld and Khattab, 1968; Baker and Rosenkrantz, 1976). Typically, maximum physiological pressure and deep body temperature are maintained (Boyde, 1976; Albrecht and Wetzel, 1979). Glauert (1975) recommends that the reservoir of a gravity-fed system should be held 120–150 cm or 20–30 cm for intra-arterial or intravenous routes, respectively.

Following anaesthesia, the animal's heart is exposed; the cannula introduced into the aorta via the left ventricle; the right atrium cut; and the preperfusion wash initiated (see below) and continued until blood is removed. The washing fluid is replaced by fixative held at deep body temperature and perfusion fixation continued for 15–30 min. The tissue is removed and fixation continued by immersion for ∿1 hr: Immersion fixation is conducted at 0–4°C.

Preperfusion washing simply serves to remove blood cells from the organ of interest. Saline was used by Glauert (1975) and Buss *et al.*, (1976), although Stumpf *et al.*, (1977) did not observe any artifacts when washing and fixation were simultaneously performed by the fixative. Nowell and Faulkin (1974) employed anticoagulants, the choice of one being dependent upon the target tissue. Nowell and Pawley (1980) note that simultaneous luminal and vascular perfusion are effective for removing mucus costs from lung and small intestine: The luminal perfusate must be isotonic (Matthieu *et al.*, 1978).

Significantly lower fixative concentration is used during perfusion because of the ease of fixative penetration into tissues. Karnovsky's fixative, having a concentration of 0.7% formaldehyde and 0.9% glutaraldehyde in slightly hypertonic cacodylate buffer, is useful (Schneeberger–Keeley and Karnovsky, 1968); while the fixative is hypertonic to the vasculature, it will be rendered isotonic by mixing with cell fluids. Various other individuals have evaluated the effect of perfusion on blood vessels (Clark and Glagov, 1976; Haudenschild and Gould, 1979); these references should be consulted when the vasculature, itself, is of interest. The vascular distribution of the target tissue itself has bearing on fixative concentration: Maunsbach (1966) perfused rat kidney with 0.25% glutaraldehyde, while brain was perfused with 2.5% glutaraldehyde (Schultz and Karlsson, 1965). Kalimo (1976) reported that higher concentrations may also be required to overwhelm the blood–brain barrier.

A simple, but effective, modification of vascular perfusion is micro-injection of the fixative. Abrunhosa (1972) and Waterman (1980) very successfully fixed embryos for SEM via microperfusion. It may also be that whole tissues are amenable to gentle microinjection when the area of interest is already exposed.

Buffer Wash

Several rinses of the specimen in chilled buffer serves to remove unreacted aldehyde from the system. Otherwise, the aldehyde will react with osmium tetroxide and leave a dense precipitate at the specimen surface (Trump and Ericsson *et al.*, 1965; Trump and Bulger, 1966). On the other hand, Ockleford (1975) claims that the buffer wash is redundant, and, provided that double fixation is cold, the buffer wash may be eliminated. As will be discussed, osmium and glutaraldehyde mixtures have been successfully used for fixation, with the rate of reaction between the two fixatives minimized by low ambient temperature and relatively short-duration exposures.

The duration of the buffer wash should be minimal to avoid extraction and consequently shrinkage. In comparison to other buffers, cacodylate promotes lipid extraction, and shrinkage will be apparent within 30 min of exposure (Bodian, 1970; Schiff and Gennaro, 1979a). To minimize these effects, the tissue should be agitated in several changes of chilled buffer for ∿15 min. Although it is known that glutaraldehyde induces a change in membrane structure, thereby potentially changing the tissue osmolarity (Breathnach and Martin, 1976; Demsey *et al.*, 1978), very little is known about this effect—and, thus, researchers maintain the same buffer formulation and osmolarity for washing.

Postfixation: Osmium Tetroxide

Osmium tetroxide, OsO_4, is a high-molecular weight (254.20) heavy metal fixative that very effectively stabilizes unsaturated lipids. When preceded by glutaraldehyde fixation, excellent general preservation of biological material results (Sabatini *et al.*, 1963; Machado, 1967; Millonig and Marinozzi, 1968; Schiechl, 1971). This fixation sequence is referred to as double fixation. Unfortunately, many researchers are not employing postosmication for SEM studies (*e.g.*, Liepins and de Harven, 1978), which is unfortunate. Extensive TEM research has confirmed that postfixation significantly enhances overall preservation: Although the SEM may not yet routinely resolve the artifacts inherent in poorly preserved tissues, standards must be maintained—*i.e.*, optimum preservation should be the standard. Furthermore, because shrinkage of biological material during processing is a well-known problem in SEM (particularly during drying), every effort should be made prior to drying to minimize shrinkage. Extraction of macromolecules during chemical fixation and dehydration inevitably induces shrinkage, but by fixing more of the cell the degree of extraction is lessened.

In addition to its qualities as a fixative, osmium tetroxide will also increase the electron density of the specimen. A number of methods have been developed which take advantage of osmium's conduction abilities; by enhancing tissue reactivity toward the metal using ligands (*e.g.*, tannin or thiocarbohydrazide), tissue density can be increased up to a point where an additional conductive coating is unnecessary. This method of increasing conductivity is discussed in Chapter 10.

Osmium will react with a variety of macromolecules (see reviews of Millonig and Marinozzi, 1968; Hayat, 1970; and Riemersma, 1970), but is most noted for its reaction with unsaturated lipids. It oxidizes unsaturated lipids at the rate of one atom of osmium per double bond (Criegee, 1936 and 1938; Criegee *et al.*, 1942; Stoeckinius and Mahr, 1965):

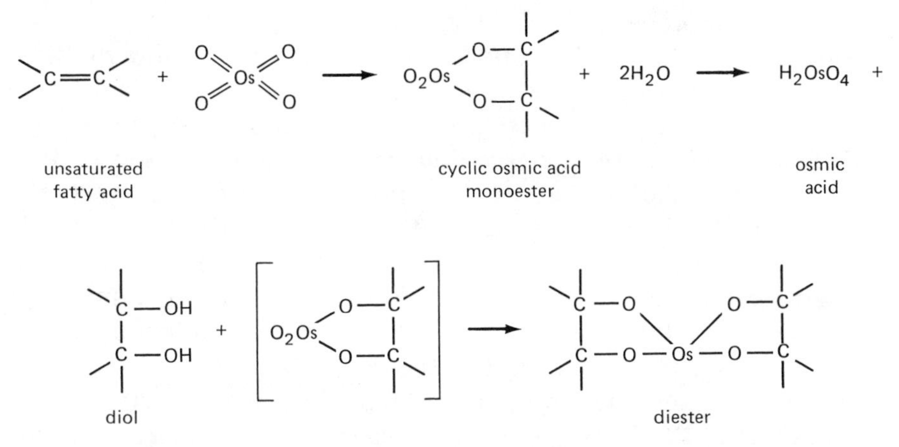

The diester product is sufficiently stable to resist dehydration (Ashworth *et al.*, 1966). The detailed chemistry of this stabilization may be found in Casley–Smith (1967), Riemersma (1968), Stein and Stein (1971), Litman and Barnett (1972), Collins *et al.* (1974a,b), and White *et al.* (1976). Saturated lipids are modified by osmium, but are extracted during dehydration (Schidlovsky, 1965). Chapman and Fluck (1966) have shown that stabilization of saturated lipids occurs only when the reaction is conducted above 60°C, but, for obvious reasons, tissue fixation is never at high temperature. Consequently, saturated lipids are extracted during tissue processing.

The reactions between osmium and proteins is a function of the protein's amino acid composition. The following bases are potential reactive sites: tryptophan (Porter and Kallman, 1953; Millonig and Marinozzi, 1968) sulfur-containing amino acids (Bahr, 1954; Adams, 1960; Hake, 1965; Elleder and Lojda, 1968a,b); and phenolic, hydroxyl, carboxyl, amino, and heterocyclic groups (Hopwood, 1969b). In comparison, Adams *et al.* (1967) and Adams and Bayliss (1968) contend that in reality the reaction is between protein-bound lipid and osmium. Elleder and Lojda (1968a) simplified the matter by showing that osmium binds to proteins, thus stabilizing them—but trace amounts are involved. They employed the OTAN (osmium tetroxide-α-naphthylamine) reaction, which is useful for detecting trace amounts of osmium. Luft and Wood (1963) describe the extraction of proteins during and after osmication, and Nielson and Griffith (1978 and 1979) thoroughly review the reactions between osmium and proteins.

The hydrophobic lipid component of lipoproteins is more reactive toward osmium than the protein component, with the lipid layer doubling in thickness (Dreher *et al.*, 1967). This is most apparent in membranes, where the increased dimensions are a function of the reaction extent, not necessarily the osmotic effect of osmium tetroxide (Pentilla *et al.*, 1974; Mersey and McCully, 1978). Also affected are the permeability properties of the membranes (Tormey, 1965; Amsterdam and Schramm, 1966): In isolated cells, osmium may cause swelling (Millonig and Marinozzi, 1968), while Moretz *et al.* (1969a) demonstrated myelin shrinkage.

Other types of macromolecules are predominately inert toward osmium. Bahr (1954) showed that nucleic acids are unaffected, and Millonig and Marinozzi (1968) demonstrated that carbohydrates are modified but not as drastically as lipids. Some modification of glycogen occurs throughout processing, but these reactions are poorly understood (De Bruijn, 1973).

The practical factors affecting osmium fixation are similar to those for aldehydes, with two major differences: First, because aldehydes convert the cytoplasm from a colloid into a spongy network, the exact tonicity of the postfixative is not quite as critical (Bone and Ryan, 1972). [Recall, however, that even after aldehyde fixation, membrane permeability is not free, but only modified from the native state. Although the osmolarity of the fixed cell is different from that of the native cell, most researchers maintain identical tonicities during double fixation (Glauert, 1975).] Second, the rate of osmium penetration is significantly slower than aldehyde rates because a barrier of heavy metal osmium atoms is erected at the fixation front. This reduces the influx of the osmium and, if specimens are too large, will remain unfixed at their core. Consequently, the rate of osmium penetration decreases with time. (Burkl and Schiechl, 1968). To avoid this problem, thin slices (~0.5-mm thickness) of tissue are osmicated.

Other factors affecting osmication are temperature and buffer type (Hagstrom and Bahr, 1960). As during primary fixation, post fixation is conducted at 0–4°C. Caulfield (1957) proved that the rate of osmium penetration increases with temperature, but Stein and Stein (1971) proved that extraction increases and uneven fixation accompany elevated temperature. Thus, the standard method is cold postfixation.

The choice of a buffer is dictated by the same factors as those described for the aldehydes (Gil and Weibel, 1968). Identical sodium phosphate buffers may be used for both primary and postfixation (Millonig, 1961); *sym*-collidine is also applicable and has the advantage of not embrittling the specimen (Gil and Weibel, 1968). Unfortunately, *sym*-collidine is toxic and expensive. Other nontraditional osmium solvents have been periodically investigated. Epling and Sjostrand (1962) and Hobbs (1969) used concentrated osmium in carbon tetrachloride; Thurston *et al.* (1976) fixed inflated lung with osmium in fluorocarbon; and Zalokar and Erk (1977) used organic solvents for osmication. Although all of these researchers achieved preservation comparable to that of buffered osmium, the latter is most commonly used.

The concentration of OsO_4 is traditionally ~2%; too high a concentration nonlinearly increases penetration, but overfixation at the tissue surface occurs. Although this is standard, Maupin–Szamier and Pillard (1978) have shown that concentrations greater than 0.05% may destroy microfilaments, in turn altering the cell conformation. Another recent observation by Shelton and Mowczko (1978) is that double fixation may induce membrane blisters—but these do not survive dehydration or critical-point drying. Although this effect should be further investigated, 2% concentration yields good fixation within 1–2 hr.

In summary, the conditions for postfixation with osmium are as follows:

Temperature	0–4°C
Time	1–2 hr
Concentration	2%
Tissue size	~0.5 mm slices
Buffer	phosphate or collidine
pH	7.2–7.4

Osmium tetroxide is an extremely hazardous chemical. Sax (1975) warns that acute or chronic exposure to osmium causes ocular disturbances and bronchitis; contact dermatitis at low concentrations; and ulceration at high concentration. It has a chlorine-like odor and a threshold limit of 0.002 mg/m^3. Osmium is handled only in a well-ventilated fume hood with the user wearing gloves and goggles. The specimens should be held in a small vial in an ice-cooled water bath held within the fume hood.

Pure osmium is commercially available in crystalline form or in aqueous solution (4% or 10%). The crystals slowly enter into solution; the buffered fixative should be prepared one day prior to use. Because small volumes (10 ml) of aqueous osmium are rapidly diluted with buffer just prior to use, the experimenter need not worry about storing stock solutions of fixative. If storage is necessary, the fixative is placed in a clear (see below) glass-stoppered bottle, wrapped in foil, labeled, and refrigerated. However, because osmium is volatile and vapors leak from the bottle, storage is not recommended.

Any organic contamination will react with osmium and, thus, lower the efficiency of fixation. Consequently, the glassware used to prepare the working solution must be very clean. Hayat (1970) recommends the following procedure:

1. Clean the glassware with soap and water followed by concentrated nitric acid.

2. Rinse several times with distilled water and air-dry the glassware (do not dry with cloth; lint may be deposited).
3. Working in a fume hood, place the vial of osmium in the clean bottle and break its seal with a glass rod.
4. Dilute with buffer. Crystals will enter solution more rapidly if the vessel is ultrasonicated.

Postfixation is followed by several rinses of cold, distilled water over ~15 min. This removes unreduced osmium from the tissue, which potentially could react with subsequent reagents (*e.g.*, acetone) and produce a dense precipitate at the tissue surface. The wash must be rapid to avoid swelling (Pentilla *et al.*, 1975); washing is accelerated by constant agitation. If ligand-mediated binding is desired with thiocarbohydrazide, it is conducted at this point.

Because osmium tetroxide is quite expensive, methods have been developed to regenerate used osmium solutions. The reader should refer to Schlatter and Schlatter–Lanz (1971) and Kiernan (1978) for these procedures.

Simultaneous Glutaraldehyde-Osmium Tetroxide Fixation

Various degrees of success have been encountered when glutaraldehyde and osmium tetroxide are simultaneously applied for primary fixation. A mixed fixative has the advantage that extraction is diminished (Trump and Bulger, 1966), although Hopwood (1970) pointed out that the different molecules may undesirably compete with one another. For example, proteins may be blocked by osmium rather than the preferred glutaraldehyde crosslinking.

Fixation must be conducted at 0–4°C for this method to be successful. Higher temperatures promote reaction between glutaraldehyde and osmium, and even at 0–4°C the duration of fixation is ~1 hr—again, to avoid this undesired reaction (Trump and Bulger, 1966). This method has been successfully used by Trump and Bulger (1966), Hirsch and Fedorko (1968), and Franke *et al.* (1969), but the conventional double-fixation sequence is most commonly used.

Uranyl Acetate

The use of uranyl acetate as a heavy metal stain has been extensively used in TEM studies of bacteria (Ryter and Kellenberger, 1958),

DNA (Kellenberger *et al.*, 1958; Huxley and Zubay, 1961; Zobel and Beer, 1965), viruses (Valentine, 1958), and for selectively staining phage within bacteria (Sechaud and Kellenberger, 1972). With whole tissues, uranyl acetate reacts with membranes (Silva *et al.*, 1968 and 1971; Brightman and Reese, 1969; Ting–Beall, 1980). As a result of these reactions, uranyl acetate will both stabilize and enhance the electron density of nucleic acids and membranes (Hayat, 1968 and 1969; Terzakis, 1968; Mumaw and Munger, 1971); when preceded by double fixation, the entire fixative process is referred to as tertiary fixation. While the advantages of uranyl acetate have been most popular in TEM, undoubtedly uranyl acetate will find increasing application as a site-selective marker in SEM studies of isolated nuclei and other DNA-containing cell organelles. Under stringent conditions, it may become possible to isolate these reaction centers by backscattered electron imaging and x-ray mapping. A great deal more research will be necessary to define these mechanisms and applications of uranyl acetate in SEM (see, *e.g.*, Lee *et al.*, 1979).

Dilute (~1.5%) uranyl acetate is applied as an aqueous solution after the postosmication water-wash (De Petris, 1965; Terzakis, 1968), or in the first step of dehydration (*e.g.*, ~1.5% uranyl acetate in dilute acetone). When phosphate or cacodylate buffered fixatives are used, thorough rinsing of the specimen with water (following osmium fixation) is essential to avoid reaction between the buffer and uranium (Farquhar and Palade, 1965). Terzakis (1968) determined that aqueous solutions are more effective in enhancing electron density than dehydration reagent solutions, because the pH at which the reaction occurs controls whether a general or selective reaction occurs: At low pH the uranyl acetate molecule selectively stains DNA, while, at higher pH, a general increase in overall electron density occurs (Wolfe *et al.*, 1962). As a general stain, uranyl acetate reacts with the chemical species and organelles noted above, as well as enhancing the density of osmicated structures. The chemistry of these reactions is discussed by Hayat (1970), Lombardi *et al.*, (1971), and Silva (1973).

Stock solutions of uranyl acetate are unstable and rapidly degrade with time. If the solution is cloudy, discard it. It is recommended that the uranyl acetate be prepared just prior to use; because it has low solubility, ultrasonication of the solution is useful. Because the concentrations noted above are near saturation, excess uranyl acetate may be present: Let the excess naturally settle at the vessel's bottom and draw the fixative from the top, or filter just before use. Because uranium is a hazardous compound, care must be exercised when handling it (Darley and Ezoe, 1976).

ORGANIC DEHYDRATION

Organic dehydration serves to replace the water in a specimen with another fluid that is miscible with the transitional fluid employed for critical-point drying. Water is immiscible with Freon or CO_2, the most common transitional fluids. Dehydration with acetone, ethanol, or 2,2-dimethoxypropane (DMP) is done with a graded series (*e.g.*, 30, 50, 70, 85, 95, 100, 100%) of the reagent in water—to avoid osmotic shock and ensure complete displacement of water (Cohen, 1979). While the 30% and 50% steps are conducted for 5–10 min, the latter steps are increased to 10–20 min to ensure adequate infiltration. Characteristics of these reagents are discussed below; more complete reviews are available in Jalanti and Demierre (1976), Kahn *et al.* (1977), Boyde and Macconachie (1979), and Howard and Postek (1979) published a bibliography of dehydration (both organic and physical).

Some degree of distortion is inevitably introduced into any specimen when its living dimensions are compared to its dehydrated dimensions (Ward and Gloster, 1976). This is due to the extraction of unfixed, and sometimes fixed, macromolecules by the dehydration reagent, and appears as shrinkage. The degree of shrinkage is controlled by the use of a graded series and by minimizing the duration of exposure. The rate and thoroughness of dehydration may be accelerated by conducting it at room temperature, constantly agitating the specimens, and changing the same concentration of solution several times (Tormey, 1965; Benscome and Tsutsumi, 1970; Banfield, 1976). A completely different approach, currently being investigated, involves the use of more reactive organic reagents, rather than conventional dehydration reagents: As will be discussed below, DMP is more rapid than acetone or ethanol. This is a double-edged sword, however, because there appears to be a relationship between very rapid dehydration and degree of distortion.

Acetone has been used for dehydrating animal tissues (Nowell and Pawley, 1980) and cultivated cells (Collins *et al.*, 1980), but is not recommended for plants because severe extraction of lipids distorts their surfaces (Parsons, *et al.*, 1974; Cohen, 1979). Boyde and his co-workers investigated the dimensional effects of acetone on cells, and found that during exposure to low concentrations cells will swell—but begin to shrink at higher concentrations, with maximum shrinkage at 70–80% concentrations (Boyde *et al.*, 1977; Boyde and Macconachie, 1979). While acetone dehydration is relatively rapid, it has the disadvantage of embrittling the specimen.

The other traditional dehydration reagent is ethanol. A similar pattern of swelling and shrinkage occurs; Boyde and Macconachie (1979) noted swelling below 70% ethanol, while shrinkage was maximum at 80-90%. This indirectly shows that the solvent action of ethanol is less than that of acetone: In general, overall extraction of macromolecules, especially lipids, is less with ethanol than it is with acetone (Millonig, 1966; Parsons *et al.*, 1974). Cohen (1979) has demonstrated that structural lipids in plant cells are more resistant to dehydration by ethanol than acetone. Animal tissues usually are dehydrated with ethanol, even though Page and Huxley (1963) noted more shrinkage of muscle filaments with ethanol than acetone, and Johnston and Roots (1967) observed that brain cell junctions were better preserved with acetone than ethanol.

DMP was introduced by Muller and Jacks (1975; Jacks and Muller, 1975) for rapid dehydration. It will react with cellular water and produce methanol and acetone—which are responsible for dehydration. Maser and Trimble (1977) indicate that this mixture is miscible with CO_2 for critical-point drying. Kahn *et al.* (1977) indicated that DMP and ethanol dehydration yield similar results, although Boyde *et al.* (1977) and Boyde and Macconachie (1979) indicate that DMP increases extraction and shrinkage. Various specimens dehydrated with DMP are red blood cells (Bistricky and Silverberg, 1976) and plants (Lin *et al.*, 1977; Postek and Tucker, 1977; Thorpe and Harvey, 1979); Prento (1978) has used DMP for light-microscope preparations. Clearly, more research is necessary to define the effect of DMP for a wider variety of specimens.

REFERENCES

Abrunhosa, R. (1972). Microperfusion fixation of embryos for ultrastructural studies. *J. Ultrastr. Res.* **41**:176.

Adams, C. M. W. (1960). Osmium tetroxide and the Marchi method: reaction with polar and nonpolar lipids, protein, and polysaccharide. *J. Histochem. Cytochem.* **8**:262.

—— *et al.* (1967). Osmium tetroxide as a histochemical and histological reagent. *Histochemie* **9**:68.

—— and O. B. Bayliss (1968). Reappraisal of osmium tetroxide and OTAN histochemical reactions. *Histochemie* **16**:162.

Albin, T. B. (1962). Handling and toxicology. In: *Acrolein*, C. W. Smith, ed. John Wiley and Sons, New York, p. 234.

Albrecht, R. M. and B. Wetzel (1979). Ancillary methods for biological scanning electron microscopy. *SEM, Inc.* **3**:203.

Alexa, G. *et al.* (1971). Reaction of dialdehyde with functional groups in collagen. *Rev. Tech. Ind. Cuir.* **63**:5.

Altman, P. L. and D. S. Dittmer, eds. (1973). *Biological Data Handbook*, 2nd ed. Federation of American Society for Experimental Biology, Bethesda, Md.

Amsterdam, A. and M. Schramm (1966). Rapid release of the zymogen granule protein by osmium tetroxide and its retention during fixation by glutaraldehyde. *J. Cell Biol.* **29**:199.

Arborgh, P. *et al.* (1976). The osmotic effect of glutaraldehyde during fixation. A transmission electron microscopy, scanning electron microscopy, and cytochemical study. *J. Ultrastr. Res.* **56**:339.

Artvinli, S. (1975) Biochemical aspects of aldehyde fixation and a new formaldehyde fixative. *Histochem. J.* **7**:435.

Ashworth, C. T. *et al.* (1966). Hepatic intracellular osmiophilic droplets: effects of lipid solvents during tissue preparation. *J. Cell Biol.* **31**:301.

Aso, C. and Y. Aito (1962). Studies on the polymerization of bifunctional monomers. II. Polymerization of glutaraldehyde. *Macromol. Chem.* **58**:195.

Bahr, G. F. (1954). Osmium tetroxide and ruthenium tetroxide and their reactions with biologically important substances. *Exptl. Cell Res.* **7**:757.

—— (1955). Continued studies about the fixation with osmium tetroxide. *Exptl. Cell Res.* **9**:277.

Baker, J. R. (1965). The fine structure produced in cells by fixatives. *J. Roy. Micros. Soc.* **48**:115.

—— and J. M. McCrae (1966). The fine structure resulting from fixation by formaldehyde: the effects of concentration, duration, and temperature. *J. Roy. Micros. Soc.* **58**:391.

—— and H. Rosenkrantz (1976) Volumetric instillation of fixatives and inert substances into mouse lungs. *Stain Technol.* **51**:107.

Banfield, G. W. (1976). Automation in tissue processing. In: *Some Biological Techniques in Electron Microscopy*, D. F. Parsons, ed. Academic Press, New York, p. 166.

Barnes, C. D. and L. D. Etherington (1973). *Drug Doses and Laboratory Animals.* University of California Press, Berkeley.

Bennet, H. S. and J. H. Luft (1959). *s*-Collidine as a basis for buffering fixatives. *J. Biophys. Biochem. Cytol.* **6**:113.

Benscome, S. A. and V. Tsutsumi (1970). A fast method for processing biologic material for electron microscopy. *Lab Invest.* **23**:447.

Bistricky, T. and B. A. Silverberg (1976). Chemical dehydration with 2,2-dimethoxypropane for the study of red blood cells by scanning electron microscopy. *Proc. 3rd. Ann. Meet. Micros. Soc. Can.* **3**:182.

Bodian, D. (1970). An electron microscopic characterization of classes of synaptic vesicles by means of controlled aldehyde fixation. *J. Cell Biol.* **44**:115.

Bone, Q. and E. J. Denton (1971). The osmotic effects of electron microscope fixatives. *J. Cell Biol.* **49**:571.

—— and K. P. Ryan (1972). Osmolarity of osmium tetroxide and glutaraldehyde fixatives. *Histochem. J.* **4**:331.

Bowes, J. H. and C. W. Cater (1966). The reaction of glutaraldehyde with proteins and other biological materials. *J. Roy. Micros. Soc.* **85**:193.

—— and C. W. Cater (1968). The interaction of aldehydes with collagen. *Biochem. Biophys. Acta* **168**:341.

—— *et al.* (1965). Determination of formaldehyde and glutaraldehyde bound to collagen by carbon-14 assay. *J. Amer. Leather Chem. Assoc.* **60**:275.

—— and R. H. Kenten (1949). The effect of deamination and esterification on the reactivity of collagen. *J. Biochem.* **44**:142.

—— and A. H. Raistrick (1964). The action of heat and moisture on leather. V. Chemical changes in collagen and tanned collagen. *J. Amer. Leather Chem. Assoc.* **59**:201.

Boyde, A. (1975). A method for the preparation of cell surfaces hidden within bulk tissues for examination in the SEM. *IITRI/SEM* p. 295.

—— (1976). Do's and don'ts in biological specimen preparation for the SEM. *IITRI/SEM* **1**:683.

—— *et al.* (1977). Dimensional changes during specimen preparation for the scanning electron microscope. *IITRI/SEM* **1**:507.

—— and E. Macconachie (1979). Volume changes during preparation of mouse embryonic tissue for scanning electron microscopy. *Scanning* **2**:149.

—— and P. Vesely (1972). Comparison of fixation and drying procedures for preparation of some cultured cell lines in the SEM. *IITRI/SEM* p. 265.

Breathnach, A. S. and M. Martin (1976). A comparison of membrane fracture faces of fixed and unfixed glycerinated tissue. *J. Cell Sci.* **21**:437.

Brightman, M. W. and T. S. Reese (1969). Junctions between intimately opposed cell membranes in the vertebrate brain. *J. Cell Biol.* **40**:648.

Brunk, U. *et al.* (1975). SEM of in vitro cultured cells, osmotic effects during fixation. *IITRI/SEM* p. 379.

Buckley, I. K. (1973a). Studies in fixation for electron microscopy using cultured cells. *Lab. Invest.* **29**:398.

—— (1973b). The lysosomes of cultured chick embryo cells. *Lab. Invest.* **29**:411.

Burkl, W. and H. Schiechl (1968). A study of osmium tetroxide fixation. *J. Histochem. Cytochem.* **16**:157.

Buss, H. *et al.* (1976). Endothelial surfaces of renal, coronary, and cerebral arteries. *IITRI/SEM* 2:217.

Busson-Mabillot, A. (1971). Influence de la fixation chimique sur les ultrastructures. *J. Microscopie* **12**:317.

Carstensen, E. L. *et al.* (1971). Stability of cells fixed with glutaraldehyde and acrolein. *J. Cell Biol.* **50**:529.

Casley-Smith, J. R. (1967). Some observations on the electron microscopy of lipids. *J. Roy. Micros. Soc.* **87**:463.

Cater, C. W. (1963). The evaluation of aldehydes and other bifunctional compounds as crosslinking agents for collagen. *J. Soc. Leather Trades Chem.* **47**:259.

——— (1965). Further investigations into the efficiency of dialdehydes and other compounds as crosslinking agents for collagen. *J. Soc. Leather Trades Chem.* **49**:455.

Caulfield, J. B. (1957). Effects of varying the vehicle for osmium tetroxide in tissue fixation. *J. Biophys. Biochem. Cytol.* **3**:827.

Chambers, R. W. *et al.* (1968). Glutaraldehyde fixation in routine histopathology. *Arch. Path.* **85**:18.

Chapman, D. and D. J. Fluck (1966). Physical properties of phospholipids. III. Electron microscope studies of some pure fully saturated 2,3-diacyl-*dl*-phosphatidyl-ethanolamines and phosphatidyl-cholines. *J. Cell Biol.* **30**:1.

Chisalita, D. *et al.* (1971). Kinetics of dialdehyde combination with the reactive groups of collagen. *Ind. Usoara.* **18**:269.

Clark, J. M. and S. Glagov (1976). Luminal surface of distended arteries by scanning electron microscopy, eliminating configuration and technical artifacts. *Brit. J. Appl. Phys.* **57**:129.

Claude, A. (1961). Problems of fixation for electron microscopy. Results of fixation with osmium tetroxide in acid and alkaline media. *Pathol. Biol.* **9**:933.

Cohen, A. L. (1979). Critical point drying-principles and procedures. *SEM, Inc.* 2:303.

——— and M. M. Shaykr (1974). Relations between fixation and dehydration in preserving cell morphology. *Proc. 32nd Ann. EMSA Meet.*, p. 124.

Collins, R. J. *et al.* (1974a). Staining and fixation of unsaturated lipids by osmium tetroxide. *Biochim. Biophys. Acta* **354**:152.

——— *et al.* (1974b). Reaction of osmium tetroxide with alkenes, glycols, and alkynes; oxo-osmium (VI) esters and their structures. *J. Chem. Soc.* (*A*) 1094.

Collins, V. P. *et al.* (1980). Transmission and scanning electron microscopy of whole glioma cells cultured *in vitro*. *SEM, Inc.* 2:223.

Conger, A. *et al.* (1978). The effect of aldehyde fixation on selected substrates for energy metabolism and amino acids in mouse brain. *J. Histochem. Cytochem.* **26**:423.

Criegee, R. (1936). Osmiumsäure-ester als Zwischenprodukte bei Oxydationen. *Justus-Liebigs Ann. Chem.* **522**:75.

—— (1938). Organische Osmiumverbindungen. *Angew. Chem.* **51**:519.

—— *et al.* (1942). Zur Kenntnis dur Organischen Osmiumverbindungen. *Justus-Liebigs Ann. Chem.* **550**:99.

Czarnecki, C. M. (1971). The effect of fixation on the chemical extraction of glycogen from rat liver. *Histochem. J.* **3**:163.

Darley, J. J. and H. Ezoe (1976). Potential hazards of uranium and its compounds in electron microscopy: a brief review. *J. Micros.* **106**:85.

Davey, D. F. (1973). The effect of fixative tonicity on the myosin filament lattice volume of frog muscle following exposure to normal or hypertonic Ringer. *Histochem. J.* **5**:87.

Dawson, R. M. C. *et al.* (1969). *Data for Biochemical Research*, 2nd ed. Clarendon Press, Oxford.

DeBault, L. E. (1973). A critical point drying technique for SEM of tissue culture cells grown on plastic substratum. *IITRI/SEM* p. 318.

DeBruijn, W. C. (1973). Glycogen, its chemical and morphologic appearance in the electron microscope. I. A modified osmium tetroxide fixative which selectively contrasts glycogen. *J. Ultrastr. Res.* **42**:29.

DeJong, D. W. *et al.* (1967). Glutaraldehyde activation of nuclear phosphatase in cultured plant cells. *Science* **155**:1672.

Demsey, A. *et al.* (1978). Cell surface membrane organization revealed by freeze-drying. *J. Ultrastr. Res.* **62**:13.

DePetris, S. (1965). Ultrastructure of the cell wall of *Escherichia coli*. *J. Ultrastr. Res.* **12**:247.

Deutsch, K. and H. Hillman (1977). The effect of six fixatives on the areas of rabbit neurons and rat cerebral slices. *J. Micros.* **109**:303.

Dreher, K. D. *et al.* (1967). The stability and structure of mixed lipid monolayers and bilayers. *J. Ultrastr. Res.* **19**:586.

Drier, T. M. and E. L. Thurston (1978). Preparation of aquatic bacteria for enumeration by scanning electron microscopy. *SEM, Inc.* **2**:843.

Ealker, J. F. (1964). *Formaldehyde*, 3rd ed. Chapman and Hall, London.

Eisenstat, L. F. *et al.* (1976). A technique for removing mucus and debris from mucosal surfaces. *IITRI/SEM* **2**:263.

Elleder, M. and Z. Lojda (1968a). Remarks on the detection of osmium derivatives in tissue sections. *Histochemie.* **13**:276.

—— and Z. Lojda (1968b). Remarks on the "OTAN" reaction. *Histochemie* **14**:47.

Epling, G. P. and F. S. Sjostrand (1962). *In vitro* and *in vivo* fixation of rabbit retina for electron microscopy. *Proc. 5th Int. Cong. EM*, P-5.

Ericsson, J. L. E. and P. Biberfeld (1967). Studies on aldehyde fixation: fixation rates and their relation to fine structure and some histochemical reactions in liver. *Lab. Invest.* **17**:281.

—— *et al.* (1965). Electron microscopic observations of the influence of different fixatives on the appearance of cellular ultrastructure. *Z. Zellforsch.* **66**:161.

Eyring, E. J. and J. Ofengand (1967). Reaction of formaldehyde with heterocyclic imino nitrogen of purine and pyrimidine nucleosides. *Biochemistry* **6**:2500.

Fahimi, F. D. and P. Drochmans (1965a). Essais de standardization de la fixation au glutaraldehyde. I. Purification et determination de la concentration der glutaraldehyde. *J. Microscopie* **7**:725.

—— and P. Drochmans (1965b). Essais de standardization de la fixation au glutaraldehyde. II. Influence des concentrations en aldehyde et de l'osmolalite. *J. Microscopie* **4**:737.

Falk, R. H. (1980). Preparation of plant tissues for SEM. *SEM, Inc.* **2**:79.

—— *et al.* (1971). The effects of various fixation schedules on the scanning electron microscope image of *Tropaeolum majus. Amer. J. Bot.* **58**:676.

Farquhar, M. G. and G. E. Palade (1965). Cell junctions in amphibian skin. *J. Cell Biol.* **26**:263.

Feder, N. and T. P. O'Brien (1968). Plant microtechnique: some principles and new methods. *Am. J. Bot.* **55**:123.

Flitney, E. W. (1966). The time course of fixation by formaldehyde, glutaraldehyde, acrolein, and other higher aldehydes. *J. Roy. Micros. Soc.* **85**:353.

Flood, P. R. (1975). Dry fracturing techniques for the study of soft internal biological tissues in the scanning electron microscope. *IITRI/SEM* p. 287.

Forbes, M. and N. Sperelakis (1971). Ultrastructure of lizard ventricular muscle. *J. Ultrastr. Res.* **34**:439.

Franke, W. W. *et al.* (1969). Simultaneous glutaraldehyde–osmium tetroxide fixation with postosmication. *Histochemie* **19**:162.

Gale, J. B. (1977). Differential effects of fixatives, buffers, and ionic species on the ultrastructure of heart mitochondria from resting and exhausted rats. *J. Electron Micros.* **26**:185.

Gennaro, J. F. *et al.* (1978). Reversible depletion of synaptic vesicles induced by application of high external potassium to the frog neuromuscular junction. *J. Phyiol.* (*London*) **280**:237.

Gertz, S. D. *et al.* (1975). Preparation of vascular endothelium for scanning electron microscopy: a comparison of the effects of perfusion and immersion fixation. *J. Micros.* **105**:309.

Gil, K. H. and E. R. Weibel (1968). The role of buffers in lung fixation with glutaraldehyde and OsO_4. *J. Ultrastr. Res.* **25**:331.

—— and E. R. Weibel (1969–1970). Improvements in demonstration of lining layer of lung alveoli by electron microscopy. *Resp. Physiol.* **8**:13.

Gillett, R. *et al.* (1975). Distilled glutaraldehyde: its use in an improved fixation regime for cell suspensions. *J. Micros.* **105**:325.

Glauert, A. M. (1975). Fixation, dehydration, and embedding of biological specimens. In: *Practical Methods in Electron Microscopy*, Vol. 3 Part 1, A. M. Glauert, ed. American Elsevier, New York.

—— and M. J. Thornley (1966). Glutaraldehyde fixation of Gram-negative bacteria. *J. Roy. Micros. Soc.* **85**:449.

Gomori, G. (1946). Buffers in the range of 6.5 to 9.6. *Proc. Soc. Exptl. Biol. Med.* **62**:33.

Gonzalez-Aguilar, F. (1969). Extracellular space in the rate brain after fixation with 12 *M* formaldehyde. *J. Ultrastr. Res.* **29**:76.

Good, N. E. *et al.* (1966). Hydrogen ion buffers for biological research. *Biochemistry* **5**:467.

Habeeb, A. F. S. A. and R. Hiramoto (1968). Reaction of proteins with glutaraldehyde. *Arch. Biochem. Biophys.* **126**:16.

Hagstrom, L. and G. F. Bahr (1960). Penetration rates of osmium tetroxide with different fixation vehicles. *Histochemie* **2**:1.

Hake, T. (1965). Studies on the reactions of OsO_4 and $KMnO_4$ with amino acids, peptides, and proteins. *Lab. Invest.* **14**:470.

Hampton, J. C. (1965). Effects of fixation on the morphology of Paneth cell granules. *Stain Technol.* **40**:283.

Haselkorn, R. and P. Doty (1961). The reaction of formaldehyde with polynucleotides. *J. Biol. Chem.* **236**:2738.

Haudenschild, C. C. and K. E. Gould (1979). Vascular organ culture: prevention of endothelial damage due to removal and reperfusion. *SEM, Inc.* **3**:865.

Hayat, M. A. (1968). Triple fixation for electron microscopy. *Proc. 26th Ann. EMSA Meet.*, p. 90.

—— (1969). Uranyl acetate as a stain and a fixative for heart tissue. *Proc. 27th Ann. EMSA Meet.*, p. 412.

—— (1970). *Principles and Techniques of Electron Microscopy*, Vol. 1. Van Nostrand Reinhold, New York.

—— (1978). *Introduction to Biological Scanning Electron Microscopy*, University Park Press, Baltimore.

Hirsch, J. G. and M. E. Fedorko (1968). Ultrastructure of human leukocytes after simultaneous fixation with glutaraldehyde and osmium tetroxide and "postfixation" in uranyl acetate. *J. Cell Biol.* **38**:615.

—— et al. (1968). Vesicle fusion and formation at the surface of pinocytic vacuoles in macrophages. *J. Cell Biol.* **38**:629.

Hobbs, M. J. (1969). Fixation of microscopic fresh-water green algae by 31% OsO_4 in CCl_4 in an unbuffered, two-phase fixative system. *Stain Technol.* **44**:217.

Holt, S. J. and R. M. Hicks (1961). Studies on formalin fixation for electron microscopy and cytochemical staining purposes. *J. Biophys. Biochem. Cytol.* **11**:31.

Hopwood, D. (1967a). Some aspects of fixation with glutaraldehyde: a biochemical and histochemical comparison of the effects of formaldehyde and glutaraldehyde fixation on various enzymes and glycogen, with a note on penetration of glutaraldehyde into liver. *J. Anat.* **101**:83.

—— (1967b). The behavior of various glutaraldehydes on Sephadex G-10 and some implications for fixation. *Histochemie* **11**:289.

—— (1969a). Fixatives and fixation: a review. *Histochem. J.* **1**:323.

—— (1969b). Fixation of proteins by osmium tetroxide, potassium dichromate, and potassium permanganate. *Histochemie* **18**:250.

—— (1970a). Use of isoelectric focusing to determine the isoelectric point of bovine serum albumin after treatment with various common fixatives. *Histochem. J.* **3**:201.

—— (1970b). The reactions between formaldehyde, glutaraldehyde, osmium tetroxide, and their fixation effects on bovine serum albumin and tissue blocks. *Histochemie* **24**:50.

—— (1972). Theoretical and practical aspects of glutaraldehyde fixation. *Histochem. J.* **4**:267.

—— (1975). The reactions of glutaraldehyde with nucleic acids. *Histochem. J.* **7**:267.

—— *et al.* (1970). The reactions between glutaraldehyde and various proteins. An investigation of their kinetics. *Histochem. J.* **2**:137.

Howard, K. S. and M. T. Postek (1979). Dehydration of scanning electron microscopy specimens–a bibliography. *SEM, Inc.* **2**:892.

Humphreys, W. J. (1977). Health and safety hazards in the SEM laboratory. *IITRI/SEM* **1**:537.

Huxley, H. E. and G. Zubay (1961). Preferential staining of nucleic acid-containing structures for electron microscopy. *J. Biophys. Biochem. Cytol.* **11**:273.

Igbal, S. J. and B. S. Weakley (1974). The effects of different preparative procedures on the ultrastructure of the hamster ovary. I. Effects of various fixative solutions on ovarian oocytes and their granulosa cells. *Histochem. J.* **38**:95.

Jacks, T. J. and L. L. Muller (1975). Instant chemical dehydration for electron microscopy. *Tex. Rep. Biol. Med.* **33**:352.

Jalanti, T. and G. Demierre (1976). Chemical dehydration of microorganisms for scanning electron microscope study. *Experientia* **32**:798.

Jansen, E. F. *et al.* (1971). Cross-linking of α-chymotrypsin and other proteins by reaction with glutaraldehyde. *Arch. Biochem. Biophysics.* **144**:394.

Jard, S. *et al.* (1966). Action de divers fixateurs sur la perméabilité et l'ultrastructure de la vessie de Grenouille. *J. Microscopy* **5**:31.

Johnson, J. E., Jr. (1978). Transmission and scanning electron microscope preparations of the same cell culture. *Stain Technol.* **53**:273.

Johnston, P. V. and B. I. Roots (1967). Fixation of the central nervous system with aldehydes and its effect on the extracellular space as seen by electron microscopy. *J. Cell Sci.* **2**:377.

Johnston, W. H. *et al.* (1973). Variations in glomerular ultrastructure in rat kidneys fixed by perfusion. *J. Ultrastr. Res.* **45**:149.

Jones, D. (1969). The reaction of formaldehyde with unsaturated fatty acids during histological fixation. *Histochem. J.* **1**:459.

—— (1972). Reactions of aldehydes with unsaturated fatty acids during histological fixation. *Histochem. J.* **4**:421.

—— and G. A. Gresham (1966). Reaction of formaldehyde with unsaturated fatty acids during histological fixation. *Nature* **210**:1386.

Jones, G. J. (1974). Polymerization of glutaraldehyde at fixative pH. *J. Histochem. Cytochem.* **22**:911.

Jost, P. *et al.* (1973). Fluidity of phospholipid bilayers and membranes after exposure to osmium tetroxide and glutaraldehyde. *J. Mol. Biol.* **76**:313.

Kahn, L. E. *et al.* (1977). Comparison of ethanol and chemical dehydration methods for the study of cells in culture by scanning and transmission electron microscopy. *IITRI/SEM* **1**:501.

Kalimo, H. (1976). The role of the blood–brain barrier in perfusion fixation of the brain for electron microscopy. *Histochem. J.* **8**:1.

Kalt, M. R. and B. Tandler (1971). A study of fixation of early amphibian embryos for electron microscopy. *J. Ultrastr. Res.* **36**:633.

Karlsson, U. and R. L. Schultz (1965). Fixation of the central nervous system for electron microscopy by aldehyde perfusion. I. Preservation with aldehyde perfusates *versus* direct perfusion with osmium tetroxide with special reference to membrane and the extracellular space. *J. Ultrastr. Res.* **12**:160.

Karnovsky, M. J. (1965). A formaldehyde–glutaraldehyde fixative of high osmolarity for use in electron microscopy. *J. Cell Biol.* **27**:137A.

Kellenberger, E. A. *et al.* (1958). Electron microscopy study of DNA-containing plasmas. II. Vegatative and mature phage DNA as compared with normal bacterial nucleoids in different physiological states. *J. Biophys. Biochem. Cytol.* **4**:671.

Kiernan, J. A. (1978). Recovery of osmium tetroxide from used fixative solutions. *J. Micros.* **113**:77.

Kormendy, A. C. (1975). Microorganisms. In: *Principles and Techniques of Scanning Electron Microscopy*, Vol. 3, M. A. Hayat, ed. Van Nostrand Reinhold, New York, p. 82.

Kuhn, C. and E. H. Finke (1972). The topography of the pulmonary alveolus: scanning electron microscopy using different fixative atoms. *J. Ultrastr. Res.* **38**:161.

Kuran, H. and M. J. Olzewski (1977). Effects of some buffers on the ultrastructure, dry mass content, and radioactivity of nuclei of *Haemanthus Katharinea. Micros. Acta* **79**:69.

Kuthy, E. and Z. Csapó (1976). Peculiar artifacts after fixation with glutaraldehyde and osmium tetroxide. *J. Micros.* **107**:177.

Landis, W. J. *et al.* (1980). Use of acrolein vapors for the anhydrous preparation of bone tissue for electron microscopy. *J. Ultrastr. Res.* **70**:171.

Langenberg, W. G. (1979). Chilling of tissues before glutaraldehyde fixation preserves fragile inclusions of several plant viruses. *J. Ultrastr. Res.* **66**:120.

Larsson, L. (1975). Effects of different fixatives on the ultrastructure of the developing proximal tubule in the cat kidney. *J. Ultrastr. Res.* **51**:140.

Lawton, J. and P. J. Harris (1978). Fixation of senescing plant tissues: sclerenchymatous fibre cells from the flowering stem of a grass. *J. Micros.* **112**:307.

Ledbetter, M. C. and B. E. S. Gunning (1963). Glutaraldehyde–osmic acid fixation of plant cells. *Symp. Bot. Appl. EM*, September.

Lee, R. M. K. W. *et al.* (1979). The effects of fixation, dehydration, and critical point drying on the size of cultured smooth muscle cells. *SEM, Inc.* **3**:439.

Levy, W. A. *et al.* (1965). Method for combined ultrastructural and biochemical analysis of normal tissue. *J. Cell Biol.* **27**:119.

Liepins, A. and E. de Harven (1978). A rapid method for cell drying for scanning electron microscopy. *SEM, Inc.* 2:37.

Lin, C. H. *et al.* (1977). Rapid chemical dehydration of plant materials for light and electron microscopy with 2,2-dimethyoxypropane and 2,2-diethoxypropane. *Amer. J. Bot.* **64**:602.

Litman, R. B. and R. Barrnett (1972). The mechanism of the fixation of tissue components by osmium tetroxide via hydrogen bonding. *J. Ultrstr. Res.* **38**:63.

Lojda, A. (1965). Fixation in histochemistry. *Folia Morph.* **13**:65.

Lombardi, L. *et al.* (1971). Electron staining with uranyl acetate. Possible role of free amino groups. *J. Histochem. Cytochem.* **19**:161.

Luft, J. H. (1959). The use of acrolein as a fixative for light and electron microscopy. *Anat. Rec.* **133**:305.

—— (1972). Fixation: the current situation in chemical stabilization of biological materials. *Proc. 30th Am. EMSA Meet.*, p. 132.

—— and R. L. Wood (1963). The extraction of tissue protein during and after fixation with osmium tetroxide in various buffer systems. *J. Cell Biol.* **19**:46A.

Machado, A. B. M. (1967). Straight OsO_4 *versus* glutaraldehyde-OsO_4 in sequence as fixatives for the granular vesicles in sympathetic axons of the rat pineal body. *Stain Technol.* **42**:293.

Malhotra, S. K. (1962). Experiments on fixation for electron microscopy. I. Unbuffered osmium tetroxide. *Quart. J. Micros. Sci.* **103**:5.

Maser, M. D. *et al.* (1967). Relationships among pH, osmolarity, and concentration of fixative solution. *Stain Technol.* **42**:175.

Maser, M. M. and J. J. Trimble (1977). Rapid chemical dehydration of biological samples for scanning electron microscopy using 2,2-dimethyoxypropane. *J. Histochem. Cytochem.* **25**:247.

Mathieu, O. *et al.* (1978). Differential effect of glutaraldehyde and buffer osmolarity on cell dimensions: a study on lung tissue. *J. Ultrastr. Res.* **63**:20.

Maugel, T. K. *et al.* (1980). Specimen preparation techniques for aquatic organisms. *SEM, Inc.* 2:57.

Maunsbach, A. B. (1966). The influence of different fixatives and fixation methods on the ultrastructure of rat kidney proximal tubule cells. II. Effects of varying osmolarity, ionic strength, buffer system, and fixative concentration of glutaraldehyde solutions. *J. Ultrastr. Res.* **15**:283.

Maupin-Szamier, P. and T. D. Pillard (1978). Actin filament destruction by osmium tetroxide. *J. Cell Biol.* **77**:837.

McDowell, E. M. and B. F. Trump (1976). Histological fixatives suitable for diagnostic light and electron microscopy. *Arch. Pathol. Lab. Med.* **100**:405.

McKinley, R. G. and P. Usherwood (1978). The effects of Mg ions on the fine structure of the insect neuromuscular junction. *J. Ultrastr. Res.* **62**:83.

Mersey, B. and M. E. McCully (1978). Monitoring of the course of fixation of plant cells. *J. Micros.* **114**:49.

Millonig, G. (1961). Advantages of a phosphate buffer for osmium tetroxide solutions in fixation. *J. Appl. Phys.* **32**:1637.

—— (1964). Model experiments on fixation and dehydration. *Proc. 6th Int. Cong. EM (Tokyo)* 2:21.

—— and V. Marinozzi (1968). Fixation and embedding in electron microscopy. In: *Advances in Optical and Electron Microscopy*, Vol. 2, R. Barer and V. E. Cosslett, eds. Academic Press, New York, p. 251.

Morel, F. M. M. *et al.* (1971). Quantitation of human red blood cell fixation by glutaraldehyde. *J. Cell Biol.* **48**:91.

Moretz, R. C. *et al.* (1969a). Use of small angle x-ray diffraction to investigate disordering of membranes during preparation for electron microscopy. I. Osmium tetroxide and potassium permanganate. *Biochim. Biophys. Acta* **193**:1.

—— *et al.* (1969b). Use of small angle x-ray diffraction to investigate disordering of membranes during preparation for electron microscopy. II. Aldehydes. *Biochim. Biophys. Acta* **193**:12.

Moore, D. J. and H. H. Mollenhauer (1969). Studies on the mechanism of glutaraldehyde stabilization of cytoplasmic membranes. *Proc. Ind. Acad. Sci.* **78**:167.

Moss, G. I. (1966). Glutaraldehyde as a fixative for plant tissues. *Protoplasma* **62**:194.

Muller, L. L. and T. J. Jacks (1975). Rapid chemical dehydration of samples for electron microscope examination. *J. Histochem. Cytochem.* **23**:107.

Mumaw, V. R. and B. L. Munger (1971). Uranyl acetate as a fixative from pH 2.0 to 8.0 *Proc. 29th Ann. EMSA Meet.*, p. 490.

Nielson, A. J. and W. P. Griffith (1978). Tissue fixation and staining with osmium tetroxide: the role of phenolic compounds. *J. Histochem. Cytochem.* **26**:138.

—— and W. P. Griffith (1979). Tissue fixation by osmium tetroxide: a possible role for proteins. *J. Histochem. Cytochem.* **27**:997.

Norton, T. N. *et al.* (1962). Studies in the histochemistry of plasmalogens: I. The effect of formalin and acrolein fixation on the plasmalogens of adrenal and brain. *J. Histochem. Cytochem.* **10**:375.

Nowell, J. A. and L. J. Faulkin (1974). Internal topography of the male reproductive system. *IITRI/SEM* p. 639.

—— and J. B. Pawley (1980). Preparation of experimental animal tissue for SEM. *SEM, Inc.* **2**:1.

O'Brien, T. P. *et al.* (1973). Coagulant and noncoagulant fixation of plant cells. *Aust. J. Biol. Sci.* **26**:1231.

Ockleford, C. D. (1975). Redundancy of washing in the preparation of biological specimens for transmission electron microscopy. *J. Micros.* **105**:193.

Page, S. G. and H. E. Huxley (1963). Filament lengths in striated muscle. *J. Cell Biol.* **19**:369.

Palade, G. E. (1952). A study of fixation for electron microscopy. *J. Exptl. Med.* **95**:285.

Palay, S. L. *et al.* (1962). Fixation of neural tissue for electron microscopy by perfusion with solutions of osmium tetroxide. *J. Cell Biol.* **12**:385.

Parsons, D. F. *et al.* (1974). A comparative survey of techniques for preparing plant surfaces for the scanning electron microscope. *J. Micros.* **101**:59.

Pentilla, A. *et al.* (1974). Influence of glutaraldehyde and/or osmium tetroxide on cell volume, ion content, mechanical stability, and membrane permeability of Ehrlich ascites tumor cells. *J. Cell Biol.* **63**:197.

—— *et al.* (1975). Effects of fixation and post fixation treatments on volume of injured cells. *J. Histochem. Cytochem.* **23**:251.

Perrachia, C. and B. S. Mittler (1972). New glutaraldehyde fixation procedures. *J. Ultrastr. Res.* **39**:57.

Pexieder, T. (1976). The role of buffer osmolarity on fixation for SEM and TEM. *Experientia* **32(6)**:806.

Pilstrom, L. and U. Nordlund (1975). The effects of temperature and concentration of the fixative on morphometry of rat liver mitochondria and rough endoplasmic reticulum. *J. Ultrastr. Res.* **50**:33.

Pladellorens, M. and J. A. Subirana (1975). Preservation of membrane ultrastructure with aldehyde or imidate fixatives. *J. Ultrastr. Res.* **52**:243.

Porter, K. R. and F. Kallman (1953). The properties and effects of osmium tetroxide as a tissue fixative with special reference to its use for electron microscopy. *Exptl. Cell Res.* **4**:127.

Postek, M. T. and S. C. Tucker (1977). Thiocarbohydrazide binding for botanical specimens for scanning electron microscopy. A modification. *J. Micros.* **110**:71.

Prento, P. (1978). Rapid dehydration–clearing with 2,2-dimethoxypropane for paraffin embedding. *J. Histochem. Cytochem.* **26**:865.

Quiocho, F. A. *et al.* (1967). Effects of changes in some solvent parameters on carboxypeptidase A in solution and in crosslinked crystals. *Proc. Nat. Acad. Sci.* **57**:525.

Richards, F. M. and J. R. Knowles (1968). Glutaraldehyde as a protein crosslinking reagent. *J. Mol. Biol.* **37**:231.

Riemersma, J. C. (1968). Osmium tetroxide fixation of lipids for electron microscopy: a possible reaction mechanism. *Biochim. Biophys. Acta* **152**:718.

—— (1970). Chemical effects of fixation on biological specimens. In: *Some Biological Techniques in Electron Microscopy*. D. F. Parsons, ed. Academic Press, New York, p. 69.

Robertson, E. A. and R. L. Schultz (1970). The impurities in commercial glutaraldehyde and their effect on the fixation of brain. *J. Ultrastr. Res.* **30**:275.

Robertson, J. G. *et al.* (1975). The effect of fixation procedures on the electron density of polysaccharide granules in *Nocardia corallina*, *J. Ultrastr. Res.* **52**:321.

Rosene, D. L. and M. M. Mesulam (1978). Fixation variables in horseradish peroxidase neurohistochemistry. I. The effects of fixative and perfusion procedures upon enzyme activity. *J. Histochem. Cytochem.* **26**:28.

Rudman, R. *et al.* (1978). The structure of crystalline Tris: a plastic precursor buffer, and acetylcholine attenuator. *Science* **200**:531.

Ryter, A. and E. Kellenberger (1958). Etude au microscope électronique de plasma contenant de l'acid désoxyribonucleique. I. Les nucléotides des bactéries en croissance active. *Z. Naturforsch.* **136**:597.

Sabatini, D. D. *et al.* (1962). New fixatives for cytological and cytochemical studies. *Proc. 5th Int. Cong. EM (Philadelphia)* 2:1.

—— *et al.* (1963). New means of fixation for electron microscopy and histochemistry. *Anat. Rec.* **142**:274.

—— *et al.* (1964). Aldehyde fixation for morphological and enzyme histochemical studies with the electron microscope. *J. Histochem. Cytochem.* **12**:57.

Saito, T. and H. Keino (1976). Acrolein as a fixative for enzyme cytochemistry. *J. Histochem. Cytochem.* **24**:1258.

Salema, R. and I. Brandao (1973). The use of PIPES buffer in the fixation of plant cells for electron microscopy. *J. Submicr. Cytol.* **9**:79.

Sax, N. I. (1975). *Dangerous Properties of Industrial Materials*, 4th ed. Van Nostrand Reinhold, New York.

Schidlovsky, G. (1965). Contrast in multilayer systems after various fixations. *Lab. Invest.* **14**:1213.

Schiechl, H. (1971). Der chemismus der OsO_4-fixierung und sein eninfluss auf die zellstrukter. *Acta Histochem. (Suppl.)* **10**:165.

Schiff, R. and J. F. Gennaro (1979a). The influence of the buffer on maintenance of tissue lipids in specimens for scanning electron microscopy. *SEM, Inc.* **3**:449.

—— and J. F. Gennaro, Jr. (1979b). The role of the buffer in the fixation of biological specimens for transmission and scanning electron microscopy. *Scanning* **2**:135.

—— *et al.* (1976). The influence of the buffer on maintenance of tissue lipid in specimens for SEM. *in Vitro* **12(4)**:305.

Schlatter, C. and I. Schlatter-Lanz (1971). A simple method for the regeneration of used osmium tetroxide solutions. *J. Micros.* **94**:85.

Schmalbruch, H. (1980). Delayed fixation alters the pattern of intramembrane particles in mammalian muscle fibrils. *J. Ultrastr. Res.* **70**:15.

Schneeberger-Keeley, E. E., and M. J. Karnovsky (1968). The ultrastructural basis of alveolar capillary membrane permeability to peroxidase used as a tracer. *J. Cell. Biol.* **37**:793.

Schultz, R. L. and U. Karlsson (1965). Fixation of the central nervous system for electron microscopy by aldehyde perfusion. II. Effect of osmolarity, pH of perfusate, and fixative concentration. *J. Ultrastr. Res.* **12**:187.

—— and N. M. Case (1968). Microtubule loss with acrolein and bicarbonate-containing fixatives. *J. Cell Biol.* **38**:633.

Séchaud, J. and E. Kellenberger (1972). Electron microscopy of DNA-containing plasms. IV. Glutaraldehyde-uranyl acetate fixation of virus infected bacteria for thin sectioning. *J. Ultrastr. Res.* **39**:598.

Shaw, J. (1960). The mechanism of osmoregulation. In: *Comparative Biochemistry*, M. Florkin, and H. S. Mason, eds. Academic Press, New York, p. 471.

Shay, J. W. and C. Walker (1980). Introduction to cells in culture as studied by SEM. *SEM, Inc.* 2:171.

Shelton, E. and W. E. Mowczko (1978). Membrane blisters: a fixation artifact. A study of fixation for scanning electron microscopy. *Scanning* **1**:166.

Silva, M. T. (1973). Uranium salts. In: *Encyclopedia of Microscopy and Microtechnique*, P. Gray, ed. Van Nostrand Reinhold, New York, p. 585.

—— *et al.* (1968). The fixative action of uranyl acetate in electron microscopy. *Experientia* **24**:1074.

—— *et al.* (1971). Uranyl salts as fixatives for electron microscopy. Study of membrane ultrastructure and phospholipid loss in bacilli. *Biochim. Biophys. Acta* **233**:513.

Simson, J. A. V., *et al.* (1978). Morphology and cytochemistry of acinar secretory granules in normal and isoproterenol treated rat submandibular glands. *J. Micros.* **113**:185.

Sjostrand, F. S. (1956). Electron microscopy of cells and tissues. In: *Physical Techniques in Biological Research*, Vol. 3, G. Oster, and A. W. Pollister, eds. Academic Press, New York, p. 241.

Skaer, R. J. and S. Whytock (1977). The fixation of nuclei in glutaraldehyde. *J. Cell Sci.* **27**:13.

Stein, O. and Y. Stein (1971). Lipid synthesis, intracellular transport, storage, and secretion. I. Electron microscope autoradiographic study of liver after injection of tritiated palmitate or glycerol in fasted and ethanol-treated rats. *J. Cell Biol.* **33**:319.

Stoeckinius, W. and S. C. Mahr (1965). Studies on the reaction of osmium tetroxide with lipids and related compounds. *Lab. Invest.* **14**:458.

Stoner, C. D. and H. D. Sirak (1969). Osmotically induced alterations in volume and ultrastructure of mitochondria isolated from rat liver and bovine heart. *J. Cell Biol.* **43**:521.

Stumpf, W. E. *et al.* (1977). Scanning electron microscopy of the collicular recess, the collicular recess organ, and the velum medullare anterius of the rat brain. *IITRI/SEM* **2**:579.

Tahmisian, T. N. (1964). Use of the freezing point to adjust the tonicity of fixing solutions. *J. Ultrastr. Res.* **10**:182.

Terzakis, J. A. (1968). Uranyl acetate, a stain and a fixative. *J. Ultrastr. Res.* **22**:168.

Thornwaite, J. T. *et al.* (1978). The use of electronic cell volume analysis with the AMAC II to determine the optimum glutaraldehyde fixative concentration for nucleated mammalian cells. *SEM, Inc.* **2**:1123.

Thorpe, J. R. and D. M. R. Harvey (1979). Optimization and investigation of the use of 2,2-dimethoxypropane as a dehydration agent for plants in transmission electron microscopy. *J. Ultrastr. Res.* **68**:186.

Thurston, E. L. (1978). Health and safety hazards in the SEM laboratory: update 1978. *SEM, Inc.* **2**:849.

Thurston, R. Y. *et al.* (1976). Ultrastructure of lung fixed in inflation using a new osmium fluorocarbon technique. *J. Ultrastr. Res.* **56**:39.

Ting-Beall, H. P. (1980). Interactions of uranium ions with lipid bilayer membranes. *J. Micros.* **118**:221.
Tobin, T. P. (1980). The osmotic effect of glutaraldehyde fixative components. *Proc. 38th Ann. EMSA Meet.*, p. 638.
Tomimatsu, Y. *et al.* (1971). Physical chemical observations on the α-chymotrypsin-glutaraldehyde system during formation of an insoluble derivation. *J. Colloid Interface Science* **36**:51.
Tooze, J. (1964). Measurements of some cellular changes during the fixation of amphibian erythrocytes with osmium tetroxide solutions. *J. Cell Biol.* **22**:551.
Tormey, J. (1965). Artifactural localizations of ferritin in the ciliary epithelium *in vitro. J. Cell Biol.* **25**:1.
Trnavska, Z. *et al.* (1966). Certain intermediatry metabolites and the formation of fibrils from collagen solutions. *Biochim. Biophys. Acta* **126**:373.
Trump, B. F. and R. E. Bulger (1966). New ultrastructural characteristics of cells fixed in a glutaraldehyde-osmium tetroxide mixture. *Lab. Invest.* **15**:368.
—— and J. L. E. Ericsson (1965). The effect of the fixative solution on the ultrastructure of cells and tissues. A comparative analysis with particular attention to the proximal convoluted tubule of the rat kidney. *Lab. Invest.* **14**:1245.
Tyler, W. S. *et al.* (1973). The potential of SEM in studies of experimental and spontaneous diseases. *IITRI/SEM* p. 403.
Valentine, R. C. (1958). Quantitative electron staining of virus particles. *J. Roy. Micros. Soc.* **78**:26.
van Deurs, B. and J. H. Luft (1979). Effects of glutaraldehyde fixation on the structure of tight junctions. A quantitative freeze fracture apparatus. *J. Ultrastr. Res.* **68**:160.
van Duijn, P. (1961). Acrolein-Schiff, a new staining method for proteins. *J. Histochem. Cytochem.* **9**:234.
Van Harreveld, A. and F. I. Khattab (1968). Perfusion fixation with glutaraldehyde and post-fixation with osmium tetroxide for electron microscopy. *J. Cell Sci.* **3**:579.
Ward, B. J. and J. A. Gloster (1976). Lipid losses during processing of cardiac muscle for electron microscopy. *J. Micros.* **108**:41.
Waterman, R. E. (1980). Preparation of embryonic tissues for SEM. *SEM, Inc.* **2**:21.
Watson, L. P. *et al.* (1980). Preparation of microbiological specimens for scanning electron microscopy. *SEM, Inc.* **2**:45.
Weakley, B. S. (1974). A comparison of three different electron microscopical grade glutaraldehydes used to fix ovarian tissues. *J. Micros.* **101**:127.
—— (1977). How dangerous is sodium cacodylate? *J. Micros.* **109**:249.
West. J. and J. L. Mangan (1970). Effects of glutaraldehyde on the protein loss and photochemical properties of Kale chloroplasts: preliminary studies on food conversion. *Nature* **228**:466.
Wetzel, B. *et al.* (1973). The need for positive identification of leukocytes examined by SEM. *IITRI/SEM* p. 535.

White, D. L. *et al.* (1976). The chemical nature of osmium tetroxide fixation and staining of membranes by X-ray photoelectron spectroscopy. *Biochim. Biophys. Acta* **436**:577.

Williams, S. T. *et al.* (1973). Preparation of microbes for scanning electron microscopy. *IITRI/SEM* p. 735.

Willison, J. H. M. and R. Rajaraman (1977). "Large" and "small" nuclear pore complexes: the influence of glutaraldehyde. *J. Micros.* **109**:183.

Wold, F. (1967). Bifunctional reagents. *Methods in Enzymology* **11**:617.

Wolfe, S. L. *et al.* (1962). The selective staining of nucleic acids in a model system and in tissue. *Proc. 5th Int. Cong. EM (Philadelphia)* 2:0–6.

Wolman, M. (1955). Problems of fixation in cytology, histology, and histochemistry. *Int. Rev. Cytol.* **4**:79.

—— and J. Greco (1952). The effect of formaldehyde on tissue lipids and on histochemical reactions for carbonyl groups. *Stain Technol.* **29**:317.

Wood, R. L. and J. H. Luft (1963). The influence of the buffer system on fixation with osmium tetroxide. *J. Cell Biol.* **19**:83A.

—— and J. H. Luft (1965). The influence of buffer systems on fixation with osmium tetroxide. *J. Ultrastr. Res.* **12**:22.

Wrigglesworth, J. M. *et al.* (1970). Organization of mitochondrial structure as revealed by freeze etching. *Biochim. Biophys. Acta* **205**:125.

Yamamoto, I. and B. Rosario (1967). Buffered formalin for primary fixation and preserving tissue for a long time. *Proc. 25th Ann. EMSA Meet.*, p. 24.

Zalokar, M. and I. Erk (1977). Phase partition fixation and staining of Drosophila eggs. *Stain Technol.* **52**:89.

Zeikus, J. A. and H. C. Aldrich (1975). Use of hot formaldehyde fixation in processing plant-parasitic nematodes for electron microscopy. *Stain Technol.* **50**:219.

Zobel, C. R. and M. Beer (1965). The use of heavy metal salts and electron stains. *Int. Rev. Cytol.* **18**:363.

5. Critical Point Drying

Anderson (1951) introduced electron microscopists to critical point drying (CPD), which was "rediscovered" in the late 1960s for the preparation of biological specimens for SEM (Anderson, 1966). CPD has been more useful for SEM than other forms of microscopy when it has been important to prepare specimens, in particular bulky tissues, in a manner such that distortion from drying is minimized (Boyde, 1978). CPD is the only drying method that completely eliminates the effects of surface tension forces on delicate biological materials. The mechanisms of disruption as a function of these forces was discussed in Chapter 4; recall that alternatives to critical point drying are (1) drying within a controlled atmosphere from a low-surface-tension fluid and (2) freeze-drying. Both of these will be considered in later chapters, and particular methods will be compared.

THEORY

Excellent theoretical papers detailing the physics of CPD are in Bartlett and Burstyn (1975), Burstyn and Bartlett (1975) and Daniels and Alberty (1975); the following is summarized from their data.

CPD is based on the principle that if a fluid [transitional fluid (TF)] is held within a sealed container and heated, the liquid will simultaneously expand and evaporate: The density of the liquid phase decreases as the density of the gas phase increases, but the total density of the fluid (combined gas and liquid phase densities) is unchanged.

As the densities of the two fluid phases approach the same value, a phenomenon known as "critical opalescence" occurs: This may be observed as the coalescence of distinct clots of the transitional fluid, and is an amorphous coalescence of liquid and gas. As temperature increases further, the unique critical temperature of the transitional fluid will be reached; the density of the liquid phase is identical to that of the gas phase; the meniscus distinguishing the phase boundary disappears; and surface tension is zero. This is the critical point, and it is a function of the critical temperature and critical pressure of the given transitional (CPD) fluid. At a temperature above critical, no amount of pressure can recondense the gas into a liquid, while critical pressure is the minimum pressure necessary to condense a gas into a liquid just below critical temperature. When the temperature is maintained above critical, the gas phase of the transitional fluid is slowly released from the chamber and returned to atmospheric pressure. The concept that volume is independent of pressure at the critical point is a function of the ideal gas law, which inversely relates pressure P and volume of one mole, $\overline{V}$, to temperature t(K) and R (a gas constant, atm):

$$P\overline{V} = \mathrm{R}T$$

When a specimen is infiltrated with an appropriate transitional fluid and carried through that fluid's characteristic critical point, the specimen will dry without being subject to interfacial tension.

The ideal transitional fluid for the preparation of biological specimens would obviously be water, but water has extreme critical properties: The critical temperature of water is 374°C and its critical pressure is 217.7 atm. Unfortunately, these values are beyond the tolerances of commercially available critical-point drying instruments. The second-best option would be to infiltrate the specimen with a transitional fluid that is miscible with water. Koller and Bernhard (1964) substituted nitrous oxide (N_2O, T_c = 26.5°C and P_c = 71.7 atm) for water and then critical-point dried.

Lewis *et al.* (1975) have shown that only the gas, not the liquid, phase of N_2O is miscible with water: The specimens are not infiltrated with nitrous oxide (which is condensed into the liquid phase under pressure) and, consequently, the specimens air-dry from water.

These authors also point out that some nitrous oxide dissolves in the water, but that complete solvent substitution is essential for successful CPD. Lewis *et al.* (1975) confirmed this result by air-drying the same specimen type from water; the conventional artifacts were observed. Consequently, nitrous oxide has been rejected as a transitional fluid for the drying of biological specimens. The final option, which is the standard method for CPD of biological materials, is to substitute cell water with a series of reagents that are miscible with both water and transitional fluids having critical properties within lower pressure and temperature ranges. The commonly used transitional fluids are carbon dioxide (T_c = 31.3°C, P_c = 72.9 atm) and Freon-13 ($CClF_3$; T_c = 28.9°C, P_c = 38.2 atm). Respectively, these were introduced by Anderson (1951) and Cohen *et al.* (1968). Note that these critical values are significantly lower than those of water. With either of these transitional fluids the typical sequence of specimen preparation is fixation ⟶ organic dehydration ⟶ substitution ⟶ transitional fluid infiltration ⟶ CPD. With either CPD fluid, fixation and dehydration are identical–*i.e.*, fixation with an aldehyde and post fixation with osmium; and dehydration with acetone, ethanol, or 2,2-dimethoxypropane. The intermediate fluids miscible with Freon-13 and CO_2 are acetone, amyl acetate, and Freon TF; ethanol is miscible with CO_2, but not fully with Freon-13. It is essential that the specimen be completely infiltrated with the CPD fluid, and this depends upon thorough infiltration with the intermediate fluid. Consequently, absolute miscibility is essential.

The most common transitional fluid in use today is carbon dioxide, originally introduced by Anderson (1951). Because CO_2 is a gas at atmospheric pressure, it is necessary to infiltrate the specimen within the CPD specimen chamber under pressure (at or below room temperature). The most common intermediate fluids, and their routes leading to CO_2-drying, are as follows (Lewis *et al.*, 1975; Cohen, 1977):

1. Acetone serves as both the dehydration and infiltration reagent, and is substituted with CO_2 in the CPD apparatus under pressure (Smith and Fink, 1972).
2. Ethanol is miscible with carbon dioxide and may be used as in option 1, above (De Bault, 1973).

3. DMP is miscible with CO_2 and may be used as in option 1 (Maser and Trimble, 1976; Lin *et al.*, 1977).
4. Following organic dehydration, amyl acetate in a graded series (*e.g.*, 1:1 amyl acetate:acetone, DMP, or ethanol; 100%, 100% amyl acetate) is used as the intermediate fluid. The amyl acetate is substituted with CO_2 in the CPD apparatus under pressure (Anderson, 1966). The advantage of using amyl acetate is that it's distinctive banana odor may be readily detected in low concentration: When the odor is absent, the specimen has been infiltrated with liquid CO_2. Also, its low volatility prevents specimen drying during transfer to the CPD apparatus.
5. Freon TF is miscible with acetone or ethanol and CO_2, and is used instead of amyl acetate, as in option 3 (Cohen *et al.*, 1968). The advantage of substitution with Freon TF is that it is more volatile than other intermediate fluids and trace amounts due to insufficient flushing of the specimen with liquid CO_2 do not interfere with CPD (*cf.* water and nitrous oxide; Cohen, *et al.*, 1968).

Boyde and his colleagues have extensively studied the effects of these different options on the degree of shrinkage inherently produced by ethanol or Freon-113 infiltration, and CO_2 drying (Boyde *et al.*, 1977; Boyde and Macconachie, 1979; Boyde and Boyde, 1980). They noted that shrinkage occurs during critical-point drying, as well as after removal of the specimens from the dryer; this latter effect is probably due to evaporation in air of residual reagents (*e.g.*, dehydration reagent or intermediate fluid). Boyde and Boyde (1980) determined the following: When ethanol served the dual function of dehydration reagent and intermediate fluid, liver tissue was 45.65% of its fixed volume immediately after CPD; after 25 days, the post-CPD volume was 38.86% of the original volume. When ethanol was used for dehydration, but substituted with Freon TF as the intermediate fluid, the respective fluid volumes were 50.49% and 46.36%. This data should be taken in the context that ethanol dehydration shrinks tissue an average of 26% (Eins and Wilhelms, 1976; Boyde and Macconachie, 1979). Brain and potato specimens exhibited the same pattern of shrinkage, except that plant specimens were more tolerant to CPD (in ethanol, the specimens retained 69.1% of their volume; after CPD, they retained 68.9% of their volume) (Boyde and Boyde, 1980). Similar shrinkage was observed by Lee *et al.* (1979) on cultured muscle cells. On the basis of this data, Boyde and his co-workers recommend that Freon TF be used as an intermediate fluid when specimens are to be critical-point dried from CO_2. CO_2 is more popular for CPD than Freon-13 (Boyde and Boyde, 1980), because CO_2 is inexpensive, relatively nontoxic, and readily available. The toxicity of intermediate and transitional fluids will be discussed.

Freon-13, $CClF_3$, was introduced by Cohen *et al.* (1968) as an alternative to CO_2. Its distinct advantage is that it is liquid at room temperature and atmospheric pressure, and, thus, that the specimens may be infiltrated with Freon-13 under normal working conditions. Its critical pressure (38.2 atm) and temperature (28.9°C) are also lower than the values for carbon dioxide, and, thus, may be marginally safer to use than CO_2. (Safety measures will be considered when the CPD apparatus is discussed.) Other fluorocarbons, such as Freon-23 (CHF_3) and Freon-116 are not in common use (Cohen, 1979). The intermediate fluids miscible with Freon-13 are acetone, amyl acetate, and Freon TF; ethanol is not fully miscible with Freon-13 and, consequently, an intermediate infiltration with amyl acetate or Freon TF is necessary (Lewis *et al.*, 1975). Using graded series during dehydration, and intermediate and transitional fluid infiltrations, the following options are available (Cohen *et al.*, 1968):

1. Acetone or DMP ⟶ acetone or DMP/Freon-13 ⟶ Freon-13 ⟶ CPD.
2. Ethanol ⟶ ethanol/amyl acetate ⟶ Freon TF ⟶ amyl acetate/Freon-13 ⟶ Freon-13 ⟶ CPD.
3. Ethanol ⟶ ethanol/Freon TF ⟶ Freon TF ⟶ Freon TF/Freon-13 ⟶ Freon-13 ⟶ CPD.
4. The user may modify option 1, above, and have an intermediate infiltration with Freon TF–which, again, does not interfere with CPD if trace amounts remain.

Shrinkage patterns similar to those observed in CPD with carbon dioxide have been observed (Boyde *et al.*, 1977). Despite these large changes in volume, the reader must understand that the degree of shrinkage is greatly affected by treatments prior to CPD, *i.e.*, CPD alone is not responsible for shrinkage: It is also induced during fixation and dehydration (*e.g.*, Cohen and Shaykh, 1973 and 1974; Mellor *et al.*, 1973; Parsons *et al.*, 1974; Pentilla *et al.*, 1975; Schneider, 1976; Boyde *et al.*, 1977 Cohen, 1977; Gusnard and Kirschner, 1977; Kirschner *et al.*, 1977; Billings–Gagliardi *et al.*, 1978; Schneider *et al.*, 1978; Boyde and Machonachie, 1979; Lee *et al.*, 1979; Moncur, 1979). Thus, the interrelationships among these factors must be considered when evaluating shrinkage. Boyde and Macconachie (1979) succinctly note that uniform volume changes—as occur in this process—are tolerable, provided that morphology is preserved. Gusnard and Kirschner (1977) submit data that supports uniform shrinkage of organelles and cells, rather than shrinkage from distortion of selective organelles and cells. Cohen (1977) has indicated that artifacts introduced during critical-point drying may be minimized by reducing sample size (especially when bulk tissues, *e.g.*, liver, are critical-point dried); holding the specimens in porous containers to lessen specimen damage by turbulence (when the transitional fluid is introduced into the specimen chamber and during the critical point); pausing rather than continuously flushing during infiltration of the transitional fluid; and heating slowly to just above the critical temperature (also see Boyde and Macconachie, 1979).

CPD APPARATUS

The apparatus used for CPD is quite simple—consisting basically of a sturdy metal container with a port for insertion of the specimen; entrance and exit ports for the transitional fluid; pressure and temperature gauges for monitoring conditions within the specimen chamber; and a means for cooling and heating the specimen chamber (water bath or electrical system). Figure 5-1 is based on the apparatus originally used by Anderson (1951), and commercially available models are similar to this. More modern models have been designed by Cohen *et al.* (1968), Cohen (1974), Pawley and Dole (1976), and Brown (1977).

Cohen (1977) and Humphreys (1977) reviewed the hazards of CPD: Unless properly maintained and operated the critical-point drier, also referred to as "the bomb," may explode. Most commer-

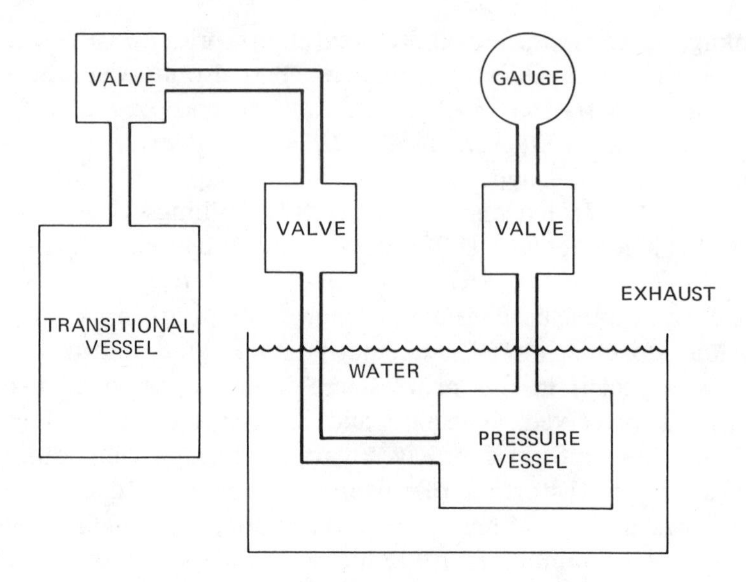

Figure 5-1. Critical-point drying apparatus.

cially available driers contain a quartz or sapphire window for visual monitoring of the events within the specimen chamber: The window must be strong enough to withstand the high pressure inherent in CPD; also, the window must be protected against damage, especially with those commercially available units in which the window is also the port for introducing samples into the specimen chamber. Carefully handle the window; if it is damaged by dropping, the user should contact the manufacturer and submit the window for strength certification; or should replace it. Similarly, any attachments to the specimen chamber must be capable of resisting high pressures. Cohen (1977) pointed out that valves, nuts, and bolts, as well as those instruments having a plexiglass safety shield over the window, must be checked regularly. The materials noted above are all potential missiles: The user must be protected by an explosion shield capable of containing high-velocity shrapnel when critical-point drying. Above all, use common sense and be careful when operating a critical-point dryer.

Another potential hazard may develop if the rubber gaskets sealing the specimen are worn or oversized, or if a noncontinuous seal resulting from extraneous material on the rubber O-ring is present. While any of these factors will probably maintain low pressure—if the pressure increases, liquid may be forced out of the specimen port at high pressure. Periodically all gaskets should be inspected and replaced when worn. Again, it is the operator's duty to maintain instrument integrity.

The temperature of the specimen chamber is maintained either electrically, by circulating water, or by submersion into water baths of appropriate temperature. The operator must know the critical pressure and temperature of the transitional fluid, and the chamber must be cooled to room temperature or lower before beginning a run. After substitution with the fluid under pressure, temperature is slowly increased to the critical point, and carried beyond that (*e.g.*, for CO_2, an increase to $\sim$ 40°C) (Cohen, 1977; Hall *et al.*, 1978): This temperature is maintained while the chamber is slowly vented and returned to atmospheric pressure. If venting is too rapid, CO_2 will freeze the exit port and decrease efficiency. Venting, particularly of fluorocarbons and amyl acetate, should be into a fume hood. Although the chronic effects of exposure to any of these chemicals, in relatively low concentrations, have not been determined, the acute effects of exposure to the fluorocarbons may include cardiac irregularity (Clayton, 1967; Deichmann and Gerarde, 1969; Silverglade, 1972; Harris, 1973; Speizer *et al.*, 1975; Humphreys, 1977). CO_2 is fatal if its concentration in air exceeds 10% (Wollman and Dripps, 1965). Fassett (1963) indicates that amyl acetate inhalation will produce intoxication, severe headache, and nausea. Consequently, all intermediate and transitional fluids should be vented into a fume hood.

A variety of specimen holders have been designed which mechanically protect fragile and/or small specimens from turbulence that may be caused during introduction of the transitional fluid or when it undergoes the phase change. Cohen (1974 and 1979) and Rosenberg (1979) thoroughly review available specimen holders; basically, a fine, porous, inert material (*e.g.*, steel mesh) may be used to hold and separate specimens. Although numerous papers exist on holders for CPD (*e.g.*, Scott *et al.*, 1973; Kurtzman *et al.*, 1974; Rice *et al.*, 1976), manufacturers now have such holders commercially available. With some ingenuity, porous but sturdy holders may be homemade. The advantages of using holders are that they allow handling of small specimens, prevent damage to fragile (*e.g.*, ciliated) structures, and avoid air-drying during transfers. Many people use holders throughout fixation, dehydration, and CPD to avoid the possibility of air-drying.

METHOD

The following serves to summarize the information presented above. Although CO_2 is the transitional fluid, the procedure below may be modified for Freon-13. Similar modifications for dehydrating in reagents other than ethanol have been discussed.

1. Dehydrate the specimen in a graded series of ethanol, up to 100% ethanol (two changes).
2. Infiltrate the specimen with a graded series of ethanol and Freon TF, up to 100% Freon TF (two changes).
3. Check the integrity of the critical-point dryer (condition of O-rings) and cool to below room temperature.
4. Place the specimens in a holder and introduce into the CPD. Take care that the specimens do not air-dry during transfer. Keep the level of Freon TF at approximately half the volume of the specimen chamber.
5. Seal the chamber, close the exit ports (gas and liquid), and introduce compressed CO_2. The pressure will rise, as will the liquid meniscus (observed through the viewing port). Release part of the liquid, reintroduce CO_2, and continue intermittent flushing for $\sim$10 min.
6. Readmit fresh CO_2 to ensure that transitional fluid infiltration is thorough. Fill the specimen chamber to approximately 75% of its capacity.
7. Slowly heat the chamber to the critical temperature (31.3°C); pressure will rise with temperature—to about 72.9 atm.
8. After passing through the critical point, maintain temperature above critical and slowly vent the specimen chamber.
9. At atmospheric pressure, remove the specimens. Examine as soon as possible, or store in a sealed container placed in a dessicator.
10. Cool the CPD.
11. Finally, always use common sense when critical-point drying.

REFERENCES

Anderson, T. F. (1951). Techniques for the preservation of 3-dimensional structures in preparing specimens for the electron microscope. *Trans. NY Acad. Sci.* **13**:130.

—— (1966). Electron microscopy of microorganisms. In: *Physical Techniques in Biological Research,* 2nd ed., Vol. 3, G. Oster, and A. Pollister, eds. Academic Press, New York, p. 319.

Bartlett, A. A. and H. P. Burstyn (1975). A review of the physics of CPD. *IITRI/SEM*, p. 305.

Billings–Gagliardi, S. *et al.* (1978). Morphological changes in isolated lymphocytes during preparation for scanning electron microscopy: freeze drying *versus* critical point drying. *Am. J. Anat.* **152**:383.

Boyde, A. (1978). Pros and cons of critical point drying and freeze drying for SEM. *SEM, Inc.* **2**:303.

—— *et al.* (1977). Dimensional changes during specimen preparation for scanning electron microscopy. *IITRI/SEM* **1**:507.

—— and S. Boyde (1980). Further studies of specimen volume changes during processing for SEM: including some plant tissue. *SEM, Inc.* **2**:117.

—— and E. Macconachie (1979). Volume changes during preparation of mouse embryonic tissue for scanning electron microscopy. *Scanning* **2**:149.

Brown, J. N. (1977). A simple low-cost critical point dryer with continuous flow dehydration attachment. *J. Micros.* **111**:351.

Burstyn, H. P. and A. A. Bartlett (1975). Critical point drying: application of the physics of the PVT surface to electron microscopy. *Am. J. Phys.* **43**:414.

Clayton, J. W., Jr. (1967). Fluorocarbon toxicity and biological action. In: *Fluorine Chemistry Reviews,* Vol. 1, P. Tarrant, ed. Marcel Dekker, Inc., New York, p. 201.

Cohen, A. L. (1974). Critical point drying. In: *Principles and Techniques of Scanning Electron Microscopy*, Vol. 1, M. A. Hayat, ed. Van Nostrand Reinhold, New York, p. 44.

—— (1977). A critical look at critical point drying-theory, practice, and artefacts. *IITRI/SEM* **1**:525.

—— (1979). Critical point drying-principles and procedures. *SEM, Inc.* **2**:303.

—— *et al.* (1968). A rapid critical point method using fluorocarbons (Freons) as intermediate and transitional fluids. *J. Micros.* **7**:331.

—— and M. Shaykh (1973). Fixation and dehydration in the preservation of surface structure in critical point drying of plant material. *IITRI/SEM,* p. 371.

—— and M. Shaykh (1974). Relations between fixation and dehydration in preserving cell morphology. *Proc. 32nd Ann. EMSA Meet.*, p. 124.

Daniels, F. and R. Alberty (1975). *Physical Chemistry,* 4th ed., Wiley–Interscience, New York.

DeBault, L. E. (1973). A critical point drying-technique for SEM of tissue culture cells grown on plastic substratum. *IITRI/SEM,* p. 317.

Diechmann, W. B. and H. W. Gerarde (1969). *Toxicology of Drugs and Chemicals.* Academic Press, New York.

Eins, S. and E. Wilhelms (1976). Assessment of preparative volume changes in central nervous tissue using automatic image analysis. *The Microscope* **24**:29.

Fassett, D. W. (1963). Esters. I. General considerations. In: *Industrial Hygiene and Toxicology,* 2nd ed., Vol. 2, D. W. Fassett and D. D. Irish, eds. Wiley Interscience, New York, p. 1847.

Gershman, H. and J. Rosen (1978). Cell adhesion and cell surface topography in aggregates of 3T3 and SV40-virus-transformed 3T3 cells. Visualization of interior cells by scanning electron microscopy. *J. Cell Biol.* **76**:639.

Gusnard, D. and R. H. Kirschner (1977). Cell and organelle shrinkage during preparation for SEM: effects of fixation, dehydration, and CPD. *J. Micros.* **110**:51.

Hall, D. J. *et al.* (1978). Critical point drying for scanning electron microscopy: a semi-automatic method of preparing biological specimens. *J. Micros.* **113**:277.

Harris, W. S. (1973). Toxic effects of aerosol propellants on the heart. *Arch. Intern. Med.* **131**:162.

Humphreys, W. J. (1977). Health and safety hazards in the scanning electron microscope laboratory. *IITRI/SEM* **1**:537.

Kirschner, R. H. *et al.* (1977). Characterization of the nuclear envelope, pore complexes, and dense lamina by high-resolution SEM. *J. Cell Biol.* **72**:118.

Koller, T. and W. Bernhard (1964). Sechage de tissus au protoxyde d'azote (N_2O) et coupe ultrafine sans matiere d'inclusion. *J. Microscopie* **3**:589.

Kurtzman, C. P. *et al.* (1974). Specimen holder to critical point dry microorganisms for scanning electron microscopy. *App. Microbiol.* **28**:708.

Lee, R. M. K. W. *et al.* (1979). The effects of fixation, dehydration, and critical point drying on the size of cultured smooth muscle cells. *SEM, Inc.* **3**:439.

Lewis, E. R. *et al.* (1975). Comparison of miscibilities and critical point drying properties of various intermediate and transitional fluids. *IITRI/SEM,* p. 317.

Lin, C. H. *et al.* (1977). Rapid chemical dehydration of plant materials for light and electron microscopy with 2, 2-dimethoxypropane and 2, 2-diethoxypropane. *Amer. J. Bot.* **64**:602.

Maser, M. D. and J. J. Trimble III (1976). Rapid chemical dehydration of mammalian tissues for scanning electron microscopy using 2,2-dimethoxypropane. *Proc. 34th Ann. EMSA,* p. 340.

Mellor, S. *et al.* (1973). Transmission electron microscopy of critical point dried tissue after observation in the scanning electron microscope. *Anat. Rec.* **176**:245.

Moncur, M. W. (1979). Shrinkage of plant material during CPD. *Scanning* **2**:175.

Parsons, E. *et al.* (1974). A comparative survey of techniques for preparing plant surfaces for the scanning electron microscopy. *J. Micros.* **101**:59.

Pawley, J. and S. Dole (1976). A totally automatic critical point dryer. *IITRI/SEM* **1**:287.

Pentilla, A. *et al.* (1975). Effects of fixation and post-fixation treatments on volume of injured cells. *J. Histochem. Cytochem.* **23**:251.

Rice, R. M. *et al.* (1976). Specimen holders for simultaneous critical point drying of multiple biological specimens. *Stain Technol.* **51**:51.

Rosenberg, W. (1979). A simple chamber allowing for increased efficiency in the handling of free cells through critical point drying: a review of techniques. *Scanning* 2:178.

Schneider, G. B. (1976). The effects of preparative procedures for scanning electron microscopy on the size of isolated lymphocytes. *Am. J. Anat.* **146**:93.

—— *et al.* (1978). Morphological changes in isolated lymphocytes during preparation for SEM: a comparative TEM/SEM study of freeze drying and critical point drying. *SEM, Inc.* 2:77.

Scott, J. R. *et al.* (1973). A fine-sieve processing container for use in CPD. *J. Micros.* **99**:359.

Silverglade, A. (1972). Cardiac toxicity of aerosol propellants. *J. Am. Med. Assoc.* **222**:827.

Smith, M. E. and E. H. Fink (1972). Critical point drying of soft biological material for the SEM. *Invest. Optham.* **11**:127.

Speizer, F. E. *et al.* (1975). Palpitation rates associated with fluorocarbon exposure in a hospital setting. *New Eng. J. Med.* **296**:624.

Wollman, H. and R. D. Dripps (1965). The therapeutic gases oxygen, carbon dioxide, and helium. In: *The Phamacological Basis of Therapeutics,* 4th ed., L. S. Goodman and A. Gilman, eds. MacMillan, New York, p. 908.

6. Freeze-Drying

A variety of cryotechniques have evolved for application in SEM—and, again, their purpose is to dry the specimen without distortion. First, specimens may be rapidly frozen at liquid nitrogen (LN_2) temperatures ($\sim$-150°C), and examined on a freezing stage within the SEM: Thus, fresh, frozen specimens are examined (Koch, 1975; Ledbetter, 1976; Echlin, 1978a,b; Fuchs *et al.*, 1978; Robards and Crosby, 1979; also see Chapter 10). Another cryotechnique is freeze-drying (FD), which involves rapid freezing of the specimen followed by sublimation of the ice under vacuum: Specimens are literally dried prior to SEM examination (Rebhun, 1972; Boyde and Echlin, 1973; Nei, 1974; MacKenzie, 1976; Coulter and Terracio, 1977; Franks, 1977). Freeze-fracturing of specimens to expose tissue interiors is also a useful method for SEM; this is usually an adjunct technique, in that it is followed by critical-point or freeze-drying (Humphreys *et al.*, 1975; Sleytr and Robards, 1977; Humphreys *et al.*, 1980). Haggis (1970) notes that the advantage of freeze-fracture is that cleavage occurs along the plane of least resistance. Alternatively, tissues may be cryofractured and observed frozen (Echlin and Burgess, 1977).

Cryotechniques are used for preparing a variety of specimens for morphological or x-ray analysis. Specimens as diverse as cartilage (Draenert and Draenert, 1979), free-living biological materials (Iwata and Aita, 1976; Schneider *et al.*, 1978), and botanical specimens (Echlin and Burgess, 1977) have been successfully freeze-dried. Numerous applications of frozen specimens in x-ray microanalysis may be found in the literature (*e.g.*, Echlin and Saubermann, 1977;

Nagy *et al.*, 1977; Saubermann *et al.*, 1977; Fuchs *et al.*, 1978; Fuchs and Fuchs, 1980; Marshall, 1980a,b). Because cathodoluminescence efficiency is inversely proportional to temperature, many autoluminescence studies of biological materials are conducted at low temperature (Horl and Ruschger, 1980).

The unique advantage of freeze-drying is that the solvent action of organic dehydration reagents is avoided, and, thus, shrinkage from extraction is lessened. This advantage is used to its fullest, for example, when it is desirable to maintain intact the waxy cuticle of plants (Cohen, 1979). Typically, freeze-drying is preceded by chemical fixation in order to maintain cell integrity; then the specimens are either rapidly frozen from water, or infiltrated with a cryoprotectant (antifreeze) and then frozen. The latter method is referred to as "freeze-substitution"; cryoprotectants such as glycerol, chloroform, and dimethyl sulfoxide (DMSO) suppress ice crystallization, resulting in a lowered level of damage from very large ice crystals (Boyde and Wood, 1969; Ellis and Mullins, 1975; Wheeler *et al.*, 1975; Echlin and Burgess, 1977; Franks *et al.*, 1977; Le B Skaer *et al.*, 1977; Barlow and Sleigh, 1979). In general, small free-living cells, such as yeast, may be frozen from water; but larger bulk specimens (*e.g.*, kidney) require cryoprotection: The larger a specimen is, the harder it is to freeze without disruptive ice crystallization (Costello, 1980). Nei (1974) indicates that a good cryoprotectant is a nonpolar solvent having a high vapor pressure below its freezing point: In practical terms, amyl acetate-infiltrated specimens completely freeze-dry, at $-75°C$ and 5×10^{-3} torr, in roughly 30 min; whereas similar conditions using hydrated specimens require up to one week. Unfortunately, freeze-substitution is effectively similar to dehydration, in that shrinkage induced by extraction occurs. When specimens are to be freeze-fractured prior to critical-point drying, they are infiltrated with the intermediate fluid and frozen at this stage: The fractured, frozen specimens are then transferred to the CPD apparatus, thawed, and infiltrated with the transitional fluid (Boyde, 1974).

Ice crystallization and the formation of large aggregates and, thus, damage, is secondly avoided by freezing very rapidly at LN_2 temperatures. LN_2 alone is unsuitable for freezing the specimen, because when anything of lower temperature is submerged in it, violent boiling ensues and the specimen becomes insulated by a bubble of N_2 gas

(Pease, 1967). Consequently, cooling rates are much slower than expected and ice crystals will grow. This problem is avoided by freezing with a LN_2-cooled slush of fluorocarbon (*e.g.*, Freon-12, with a b.p. of -29.79°C and a m.p. of -158.0°C; or Freon-22, with respective values of -40.8°C and -160.0°C) or of hydrocarbon (*e.g.*, propane, b.p. -42.12°C, m.p. -187.1°C). Umrath (1974) designed a vacuum unit which converts LN_2 to a solid–liquid slush which avoids specimen insulation; any of these methods are applicable to SEM specimen preparation. Rebhun (1972) comprehensively describes quenching media, their relative rates of cooling, and ultimately their effect on cell morphology.

Newman and Tailby (1973), Spicer, *et al.*, (1974), Albrect and MacKenzie (1975) and Hayes and Pauley (1975) have described quenching and freeze-drying units. At their simplest, quenching units are basically two-liter Dewar flasks used to contain the LN_2, and a "castle" containing the actual quenching liquid (*e.g.*, Freon 22). The castle is cooled by direct contact with an aluminum conducting rod submerged in the LN_2 (see Spicer *et al.*, 1974, for exact dimensions). To adapt this quenching unit to a freeze-drier, the surface of the aluminum block/castle are held within a bell jar having direct connections to a vacuum system. Albrecht and MacKenzie (1975) describe a more sophisticated freeze-drier, and various models are commerically available: Although any of these units are amenable to freezing, fracturing, and/or freeze drying, the basic unit of Spicer *et al.* (1974) is used for simplicity.

The rapid freezing of any specimen follows the general pathway presented below. Normally, the specimen is chemically fixed and subsequently frozen from water or from a cryoprotectant. The specimen is directly mounted on a polished stub (Boyde and Wood, 1969); if adhesion is a problem, the stub may be coated with a water-soluble adhesive, such as albumin (Boyde and Echlin, 1973). The specimen/stub is plunged into the Freon-22 slush and is very rapidly frozen. After the specimen temperature has reached equilibrium with the slush–a function of specimen size–it may be fractured simply by applying pressure with a cooled metal probe (forceps, razor, *etc.*): The tissue will fracture along natural cleavage places (Nemanic, 1972; Boyde and Echlin, 1973; Humphreys *et al.*, 1974). While frozen, the specimen is transferred to the precooled (-60°C–-80°C) freeze-drier, and a vacuum of $\sim 10^{-2}$ torr attained. Sublimation of the solid ice is thus initiated; the specimen must remain at low temperature throughout this process.

The length of time required for complete drying is a function of specimen size, volatility of the frozen medium (water or a cryoprotectant), and temperature. The smaller the specimen, the more rapid its drying. Similarly, more volatile materials sublime more rapidly. Warming the specimen increases the vapor pressure and, thus, the sublimation rate; but collapse of the specimen may occur above ∿-60°C (Albrecht and MacKenzie, 1975; Kistler and Kellenberger, 1977). Albrecht and MacKenzie (1975) note that the best indication of complete dehydration is obtained by monitoring vapor pressure within the specimen chamber: If a slight increase of temperature is accompanied by a rise in vapor pressure, the specimen has not completely dried. When drying is complete, the chamber is slowly warmed to room temperature while under vacuum; at room temperature the specimen chamber is backfilled with an inert gas (argon or nitrogen), the specimen coated, and promptly examined in the SEM. Rehydration by exposure to atmospheric humidity must be avoided; always store specimens in a desiccator.

Humphreys *et al.* (1974, 1975, 1980) have thoroughly investigated cryofracturing of specimens that ultimately will be critical-point dried. They note that the primary advantage of freeze-fracturing is that mechanical distortion inherent in cutting or tearing of soft specimens is avoided. Because organic dehydration and substitution are prerequisites for CPD, these authors, and others (Boyde and Wood, 1969), have shown that specimens may be frozen in ethanol, acetone, amyl acetate, or Freon TF; fractured; returned to room temperature; and critical-point dried without disruption: This has been confirmed by TEM (Humphreys, 1975; Humphreys *et al.*, 1975). Following conventional fixation and dehydration with ethanol, Humphreys *et al.* (1978) used this method for cryofracturing bulk tissues:

1. In 100% ethanol, the specimen (which is cut in strips) is slipped into a Parafilm sheath, and the Parafilm is sealed at both ends. The specimen is thus secured while still bathed in ethanol.
2. Grasping the sheath with forceps, it is frozen in LN_2; then transferred to a LN_2-precooled metal block and fractured with a LN_2-cooled razor blade.
3. The Parafilm sheath is submerged in fresh ethanol at room tem-

perature, the sheath removed, and the specimens critical point dried.

4. Following CPD, fracture faces are distinguished from other planes by their smooth and shiny appearance.

The methods discussed above have all been successfully used in SEM. The theoretical discussion of events during the freezing process have been ignored primarily because critical-point drying is far more popular than freeze-drying when the ultimate goal is examination of morphological features. On the other hand, freezing methodologies are almost exclusively used when elemental microanalysis is desired (see earlier reference, especially Marshall, 1980a,b). Excellent theoretical descriptions of freezing are found in Rebhun (1972) and Nermut (1977).

COMPARISON OF FD AND CPD

Biological specimens will shrink during any preparation: on the average, freeze-dried specimens undergo ∿15% volume reduction, whereas critical point dried specimens are reduced to ∿40% of their native volume (Boyde *et al.*, 1977; Boyde, 1978). In the latter case, the solvent influence of organic dehydration must be considered—*i.e.*, shrinkage is not solely a function of CPD (Boyde and Macconachie, 1979). When a specimen is of uniform composition, shrinkage in all probability is equal throughout the bulk (Gusnard and Kirschner, 1977). In comparison, multiphase specimens, such as cartilage and bone junctions (Draenert and Draenert, 1979), or monolayer cultures (Albrecht and MacKenzie, 1975) present special problems (Nagy *et al.*, 1977). In the bone specimen, for example, the water of hydration is different from one morphologic area to the next, and ice crystallization damage is only reduced by freezing at, or lower than, -90°C (Draenert and Draenert, 1979).

In short, the experimenter must tolerate the artifacts induced during any drying method, and whenever possible correlate data with that obtained by light or transmission electron microscopy. If the instrumentation is available, the experimenter should compare the effects of different preparatory procedures and always consult the literature (*e.g.*, Cole and Ramirez–Mitchell, 1974; Schneider, 1976; Billings–Gagliardi, *et al.*, 1978; Boyde, 1978; Schneider *et al.*, 1978).

REFERENCES

Albrecht, R. M. and A. P. MacKenzie (1975). Cultured and free-living cells. In: *Principles and Techniques of Scanning Electron Microscopy*, Vol. 3, M. A. Hayat, ed. Van Nostrand Reinhold, New York, p. 109.

Barlow, D. I. and M. A. Sleigh (1979). Freeze substitution for preservation of ciliated surfaces for scanning electron microscopy. *J. Micros.* **115**:81.

Billings-Gagliardi, S. *et al.* (1978). Morphological changes in isolated lymphocytes during preparation for scanning electron microscopy. Freeze drying *versus* critical point drying. *Am. J. Anat.* **152**:383.

Boyde, A. (1974). Freezing, freeze fracturing, and freeze drying in biological specimen preparation for the SEM. *IITRI/SEM*, p. 1044.

—— (1978). Pros and cons of critical point drying and freeze drying for SEM. *SEM, Inc.* **2**:303.

—— *et al.* (1977). Dimensional changes during specimen preparation for scanning electron microscopy. *IITRI/SEM* **1**:507.

—— and P. Echlin (1973). Freezing and freeze-drying–a preparative technique for SEM. *IITRI/SEM*, p. 759.

—— and E. Macconachie (1979). Volume changes during preparation of mouse embryonic tissue for scanning electron microscopy. *Scanning* **2**:149.

—— and C. Wood (1969). Preparation of animal tissues for surface scanning electron microscopy. *J. Micros.* **90**:221.

Cohen, A. L. (1979). Critical point drying–principles and procedures. *SEM, Inc.* **2**:303.

Cole, G. T., and R. Ramirez-Mitchell (1974). Comparative scanning electron microscopy of *Penicillium sp.* conidia subjected to critical point drying, freeze drying, and freeze etching. *IITRI/SEM*, p. 367.

Costello, M. J. (1980). Ultra-rapid freezing of thin biological samples. *SEM, Inc.* **2**:361.

Coulter, H. D. and L. Terracio (1977). Preparation of biological tissues for electron microscopy by freeze drying. *Anat. Rec.* **187**:477.

Draenert, S. and K. Draenert (1979). Freeze drying of articular cartilage. *Scanning* **2**:57.

Echlin, P., ed. (1978a). Low Temperature Biological Microscopy. *J. Micros.*, (Oxford) **111–112**.

—— (1978b). Low temperature scanning electron microscopy: a review. *J. Micros.* **112**:47.

—— and A. Burgess (1977). Cryofracturing and low temperature scanning electron microscopy of plant material. *IITRI/SEM* **1**:491.

—— and A. J. Saubermann (1977). Preparation of biological specimens for x-ray microanalysis. *IITRI/SEM* **1**:621.

—— *et al.* (1977). Polymeric cryoprotectants in the preservation of biological ultrastructure. II. Physiological effects. *J. Micros.* **110**:239.

Ellis, E. A., and J. T. Mullins (1975). Preparation of coenocytes for freeze etching. *Stain Technol.* **50**:245.

Franks, F. (1977). Biological freezing and cryofixation. *J. Micros.* **111**:1.

—— *et al.* (1977). Polymeric cryoprotectants in the preservation of biological ultrastructure. I. Low temperature states of aqueous solutions. *J. Micros.* **110**:223.

Fuchs, W. and H. Fuchs (1980). The use of frozen-hydrated bulk specimens for x-ray microanalysis. *SEM, Inc.* **2**:371.

—— *et al.* (1978). Instrumentation and specimen preparation for electron beam x-ray microanalysis of frozen hydrated bulk tissue. *J. Micros.* **112**:75.

Gusnard, D. and R. H. Kirschner (1977). Cell and organelle shrinkage during preparation for SEM: effects of fixation, dehydration, and CPD. *J. Micros.* **110**:51.

Haggis, G. H. (1970). Cryofracture of biological material. *IITRI/SEM*, p. 97.

Hayes, T. L. and J. B. Pawley (1975). Very small biological specimens. In: *Principles and Techniques of Scanning Electron Microscopy*, Vol. 3, M. A. Hayat, ed. Van Nostrand Reinhold, New York, p. 45.

Horl, E. M. and P. Roschger (1980). CL SEM investigations of biological material at liquid helium and liquid nitrogen temperatures. *SEM, Inc.* **1**:285.

Humphreys, W. J. (1975). Drying soft biological tissues for scanning electron microscopy. *IITRI/SEM*, p. 707.

—— *et al.* (1974). Critical point drying of ethanol infiltrated, cryofractured, biological specimens for scanning electron microscopy. *IITRI/SEM*, p. 275.

—— *et al.* (1975). Transmission electron microscopy of tissue prepared for scanning electron microscopy by ethanol-cryofracturing. *Stain Technol.* **50**:119.

—— *et al.* (1980). Critical point drying of ethanol infiltrated, cryofractured, biological specimens for scanning electron microscopy. *IITRI/SEM*, p. 275.

Iwata, H., and S. Aita (1976). Freeze drying technique for small biological objects. *J. Electron Micros.* **25**:205.

Kistler, J. and E. Kellenberger (1977). Collapse phenomena in freeze drying. *J. Ultrastr. Res.* **59**:70.

Koch, G. R. (1975). Preparation and examination of specimens at low temperature. In: *Principles and Techniques of Scanning Electron Microscopy*, Vol. 4, M. A. Hayat, ed. Van Nostrand Reinhold, New York, p. 1.

Le B Skaer, H., *et al.* (1977). Polymeric cryoprotectants in the preservation of biological ultrastructure. III. Morphological structures. *J. Micros.* **10**:257.

Ledbetter, M. C. (1976). Practical problems in observation of unfixed, uncoated plant surfaces by SEM. *IITRI/SEM* **2**:453.

MacKenzie, A. P. (1976). Principles in freeze drying. *Transp. Proc.* **8**:181.

Marshall, A. T. (1980a). Freeze substitution as a preparation technique for biological x-ray microanalysis. *SEM, Inc.* **2**:395.

—— (1980b). Quantitative x-ray microanalysis of frozen-hydrated bulk biological specimens. *SEM, Inc.* **2**:349.

Nagy, I. Z. *et al.* (1977). Energy-dispersive x-ray microanalysis of electrolytes in biological specimens. I. Specimen preparation, beam penetration, and quantitative analysis. *J. Ultrastr. Res.* **58**:22.

Nei, T. (1974). Cryotechniques. In: *Principles and Techniques of Scanning*

Electron Microscopy, Vol. 1, M. A. Hayat, ed. Van Nostrand Reinhold, New York, p. 113.

Nemanic, M. K. (1972). Critical point drying, cryofracture, and serial sectioning. *IITRI/SEM*, p. 297.

Nermut, M. V. (1977). Freeze drying for electron microscopy. In: *Principles and Techniques of Electron Microscopy*, Vol. 7, M. A. Hayat, ed. Van Nostrand Reinhold, New York, p. 79.

Newman, H. N. and P. W. Tailby (1973). A simple apparatus for the freeze drying of biological specimens preparatory to scanning electron microscopy. *Stain Technol.* **48**:145.

Pease, D. C. (1967). Eutectic ethylene glycol and pure propylene glycol as substituting media for the dehydration of frozen tissue. *J. Ultrastr. Res.* **21**:75.

Rebhun, L. I. (1972). Freeze-substitution and freeze-drying. In: *Principles and Techniques of Electron Microscopy*, Vol. 2, M. A. Hayat, ed. Van Nostrand Reinhold, New York, p. 3.

Robards, A. W. and P. Crosby (1979). A comprehensive freezing, fracturing and coating system for low temperature scanning electron microscopy. *SEM, Inc.* 2:325.

Saubermann, A. J. *et al.* (1977). Preparation of frozen hydrated tissue sections for x-ray microanalysis using a satellite vacuum coating and transfer system. *IITRI/SEM* **1**:347.

Schneider, G. B. (1976). The effects of preparative procedures for scanning electron microscopy on the size of isolated lymphocytes. *Am. J. Anat.* **146**:93.

—— *et al.* (1978). Morphological changes in isolated lymphocytes during preparation for SEM: a comparative TEM/SEM study of freeze-drying and critical point drying. *SEM, Inc.* **2**:77.

Sleytr, U. B. and A. W. Robards (1977). Freeze fracture: a review of methods and results. *J. Micros.* **111**:77.

Spicer, R. A. *et al.* (1974). An inexpensive portable freeze drying unit for SEM specimen preparation. *IITRI/SEM*, p. 299.

Umrath, W. (1974). Cooling bath for rapid freezing in electron microscopy. *J. Micros.* **101**:103.

Wheeler, E. E. *et al.* (1975). Freeze drying from tertiary butanol in the preparation of endocardium for scanning electron microscopy. *Stain Technol.* **50**:331.

7. Specimen Drying from Volatile Reagents

Chemical fixation and dehydryation methods function to bring a well-preserved specimen to the dry state, thus preparing it for vacuum exposure. Simple air-drying of specimens from the natural hydrated state severely disrupts structural integrity. Relatively large cells have a large volume-to-surface ratio and are most influenced by volume effects. For example, native bacteria will shrink and collapse, and plant cells shrink by plasmolysis, when air-dried from water (*i.e.*, unmodified). As organisms become smaller (*e.g.*, viruses), the volume-to-surface ratio proportionately decreases and surface forces become predominant. This destructive force is interfacial tension, which is the pressure difference (force) between two sides (gas and liquid) of a liquid meniscus: As water-suspended cells air-dry, the receding liquid meniscus passes through them and builds up enormous forces. Anderson (1951) calculated that the stress through an air-drying flagellum in an incredible 46,000 kg/cm^2.

Burstyn and Bartlett (1975) effectively described these forces as similar to a hydraulic pump: As the piston/water-vapor interface descends, everything in its path is crushed. The evaporation direction is cylindrically downward, and as evaporation proceeds the cylinder radius decreases by an inverse proportion with pressure. The amount of induced axial compression is sufficient to destroy delicate biological materials.

The conventional drying methods employed in SEM, which avoid damage by interfacial tension, are air-drying from low surface tension fluids, critical-point drying, and freeze-drying. Another option, which does not employ any drying procedure, is the examination of frozen tissues held on a cold-stage in the SEM; this method is useful for plant materials and is discussed in Chapter 10. These drying methods all have in common the goal of reducing or eliminating interfacial tension. Air-drying from low-surface-tension fluids and freeze-drying reduce interfacial tension, whereas critical-point drying completely eliminates this force. Depending upon the type of specimen, goals of the experiment, and equipment available, a choice of a given drying method is made. The following discussion of air-drying is the simplest technique, but it is not universally applicable. As will be discussed, most researchers choose critical-point or freeze-drying for optimal preservation.

The possibility of air-drying specimens without structural damage has long tempted biological microscopists. The reasons for this attraction are that preparation time would be lessened (although primary and postfixation, as well as dehydration, are necessary) and the potential hazards associated with the critical-point drying apparatus are eliminated. The author stresses that the air-drying methods discussed below are not equivalent replacements for critical-point or freeze-drying, but that some specimens may be air-dried without severe damage.

A few biological materials exist in a naturally semidehydrated state and require neither fixation or drying. For example, some bacterial spores and mycological specimens are transported by air; these specimens may be coated and examined without treatment (Watson *et al.*, 1980). Likewise, some pollen may be directly examined (Nickerson *et al.*, 1974). Insects equipped with a tough exoskeleton also fall in this category. However, it is stressed that these specimens normally exist at atmospheric pressures, not under high-vacuum conditions. Therefore, they may collapse when exposed to the SEM vacuum. If sufficient sample is available, the researcher should try the native specimen; if collapse is noted, the remainder of the specimen should be fixed and dried according to standard methods.

A number of water-borne plant and animal species having tough exoskeletons may be air-dried from water without surface damage. Single-celled organisms, such as some radiolarians and foraminiferans, may be air-dried and their shells examined (*e.g.*, Maugel *et al.*, 1980);

other aquatic organisms include some phytoplankton (Postek, 1975) and algae (Marchant, 1974). For example, a diatom frustule (exoskeleton) is composed of silicon and is sufficiently strong to tolerate air-drying. The suspension should be purified, however, and the cell fraction isolated from extraneous detritus. The detailed methodology is included in Chapter 8; basically, it involves centrifugation of the specimen—with several resuspensions in distilled water, followed by filtration onto a nuclear pore membrane, and finally sputter coating.

The vast majority of other biological specimens are hydrated and will be destroyed if air-dried from their native state. Consequently, these specimens (*e.g.*, bulk tissues, bacteria, blood cells, and cultured cells) are conventionally fixed prior to drying. In addition to stabilizing fine structure, fixation will also harden the tissue, making it strong enough, in some cases, to tolerate air-drying (Cohen, 1979). To some extent, the degree of specimen hydration will determine if it can withstand air-drying: For example, embryonic tissues are both fragile and contain a high proportion of water, and are absolutely impossible to air-dry without extreme shrinkage (Waterman, 1980). On the other hand, Costa *et al.* (1977) air-dried platelet dense bodies with little adverse affect.

The majority of air-drying studies involve drying from fluids having low surface tension. Water has a surface tension t of 73.0 dynes/cm; neither fixed nor unfixed tissues can tolerate air-drying from water. However, substitution of water with a lower inherent surface tension liquid, and air-drying from this, reduces damage. The most common reagents for air-drying all reduce surface tension by a factor of at least three: These are acetone (t = 23.7 dynes/cm), ethanol (t = 22.8 dynes/cm), and Freon TF (or Freon-113; t = 19.0 dynes/cm). Following fixation, the tissues are dehydrated and dried from 100% acetone or ethanol; when Freon-113 is used, conventional dehydration is followed by substitution with Freon-113 for evaporation. Other fluids of lower surface tension are not recommended, because too often they are polar molecules which may damage surfaces (Albrecht *et al.*, 1976).

Some of the tissues which have been successfully air-dried from these low-surface-tension fluids include botanical specimens (Parsons *et al.*, 1974); blood cells (Bessis and Weed, 1972; Polliack *et al.*, 1973 and 1974; Thornwaite and Leif, 1974; Liepins and de Harven, 1978); bacteria (Klainer *et al.*, 1974; Colvin and Leppard, 1977); tissue cultures (Liepens and de Harven, 1978; Lamb and Ingram, 1979); bulk specimens of bladder, kidney, and uterus (Lamb and Ingram, 1979); and cartilage (Hesse and Hesse, 1978).

Until recently, most researchers allowed cells to air-dry directly into the atmosphere (*e.g.*, Boyde and Vesely, 1972). More current research, however, indicates that ambient humidity plays an important role in the success rate of "air-drying." Therefore, different researchers have altered the atmosphere under which drying occurs. Albrecht and MacKenzie (1975) dried cells from ethanol or acetone in a desiccator at -30°C--40°C and had good results; Lamb and Ingram (1979) had very-well-preserved bladder, kidney, and uterus specimens dried from ethanol or Freon-113 in a desiccator (anhydrous $CaSO_4$) and under an argon atmosphere (1 psi), although trachea prepared in this manner exhibited some microvillus collapse. Going in the opposite direction (under vacuum) de Harven *et al.*, (1977) and Liepins and de Harven (1978) dried cultured cells and red blood cells under a vacuum of 10^{-2} torr from ethanol or Freon-113, with results comparable to that of critical-point drying. Some reviewers expressed the concern that the cells were freezing during vacuum exposure; but this was disproved by de Harven *et al.* (1977), who showed that at 2.3×10^{-2} torr, the temperature of evaporation was -75°C, whereas the freezing point of ethanol is -117°C. Thus, this method is referred to as "low-temperature vacuum-drying".

The method used by Liepin and de Harven (1978) for drying red blood cells and cultured cells under vacuum, is as follows:

1. The cells were fixed in 2.5% glutaraldehyde (0.1 *M* cacodylate buffer containing 0.01% $CaCl_2$) for 1 hr at 22°C.
2. The cells were rinsed in cacodylate buffer overnight.
3. Ethanol dehydration at 25, 50, 75, 100, 100% was for 4 min at each step.
4. The cells were infiltrated in a graded series of ethanol and Freon-113 (25, 50, 75, 95, 100, 100%) for 4 min each.
5. The specimens suspended in pure Freon-113 were transferred to a desiccator, excess liquid was drained, and the chamber was evacuated to 2.3×10^{-2} torr.

6. Depending upon the size of the sampling, drying was complete in 5–30 min.
7. The samples were mounted on a stub, sputter coated, and examined.

These researchers note that cells prepared in this manner were identical to controls prepared by critical-point drying. They also note that earlier attempts at air-drying of cultures (Boyde and Vesely, 1972) failed because ambient humidity probably was high; *i.e.*, the cells were slightly hydrated by exposure to atmospheric water and, consequently, air-dried from water—not a low-surface-tension fluid.

Lamb and Ingram (1979) prepared bulk tissues as follows:

1. Aldehyde fixation for different tissues was:
   a. *Trachea*: 4% glutaraldehyde with 1% formaldehyde in 0.1 *M* cacodylate buffer (Karnovsky's fixative), 2 hr.
   b. *Peritoneal macrophages and uterus*: 2.5% glutaraldehyde in phosphate buffer, 2 hr.
   c. *Kidney*: 2% glutaraldehyde in 0.05 *M* sodium cacodylate, 2 hr.
   d. *Bladder*: 1.3% glutaraldehyde in 0.05 *M* sodium cacodylate, 2 hr.
2. The specimens were dehydrated in graded series (35, 50, 70, 95, 100, 100, 100%) of ethanol for 15 min each.
3. Various drying routes were used for comparison:
   a. Air drying at ∿21°C, relative humidity ∿50% always resulted in collapse.
   b. The Liepin and de Harven (1978) method was used successfully; Lamb and Ingram (1979) noted that vacuum-drying was effective from either ethanol or Freon TF.
   c. Drying in an argon atmosphere in a desiccator (all ports containing anhydrous $CaSO_4$) was successful, although some ciliated areas were collapsed in the trachea specimen. Drying of individual cells was complete within 10 min, and bulk tissue within 40 min.
   d. Critical-point dried specimens were used as controls.

Both the Liepin and de Harven (1978) and Lamb and Ingram (1979) methods deserve consideration as useful alternatives to critical-point or freeze-drying: These methods are not intended as replacements for more conventional drying methods, but may be useful in routine situations. More experimentation will be necessary before argon- or vacuum-drying can attain respectability and, thus, acceptance; for example, neither group of researchers used osmium tetroxide postfixation, although they state that unpublished results indicate that postosmication has no effect on external structural features. The influence of different fixatives may actually be a double-edged sword. Whereas double fixation is the standard for ultrastructural study alone, the tissue is rendered brittle. On the other hand, tissue fixed solely in aldehyde remains somewhat elastic. In may be that either method strengthens the tissue by an increase in native density (*i.e.*, double fixation) and/or by increasing elasticity and, therefore, ability to rebound from stress. If an increase in density is, indeed, one mode of physical stabilization, the effects of air-drying on specimens which have undergone metallic impregnation needs investigation. Further experiments should evaluate volume effects (*i.e.*, degree of shrinkage relative to other drying methods) and confirm ultrastructural integrity by TEM examinations. Clearly, because Lamb and Ingram (1979) have successfully prepared both single-cell and bulk-tissue specimens that exhibit minimal deformation as compared to the severe deformation normally inherent in air-dried bulk specimens, it is probable that many more specimen types may be simply prepared.

REFERENCES

Albrecht, R. M. and A. P. MacKenzie (1975). Cultured and free living cells. In: *Principles and Techniques of Scanning Electron Microscopy*, Vol. 3, M. A. Hayat, ed. Van Nostrand Reinhold, New York, p. 109.

—— *et al.* (1976). Preparation of cultured cells for SEM: air-drying from organic solvents. *J. Micros.* **108**:21.

Anderson, T. F. (1951). Techniques for the preservation of 3-dimensional structures in preparing specimens for the electron microscope. *Trans. N. Y. Acad. Sci.* **13**:130.

Bessis, M. and R. I. Weed (1972). Preparation of red blood cells (RBC) for SEM. A survey of various artifacts. *IITRI/SEM* p. 289.

Boyde, A. and P. Vesely (1972). Comparison of fixation and drying procedures for preparation of some cultured cell lines for examination in the SEM. *IITRI/SEM* p. 265.

Burstyn, H. P. and A. A. Bartlett (1975). Critical point drying: application of the physics of the PVT surface to electron microscopy. *Am. J. Phys.* **43(5)**:414.

Cohen, A. L. (1979). Critical point-drying–principles and procedures. *SEM, Inc.* 2:303.

Colvin, J. R. and G. G. Leppard (1977). The biosynthesis of cellulose by *Acetobacter xylinum* and *Acetobacter acetigenus. Can. J. Microbiol.* **23**:701.

Costa, J. L. *et al.* (1977). Evaluation of the utility of air dried whole mounts for quantitative electron microprobe studies of platelet dense bodies. *J. Histochem. Cytochem.* **25**:1079.

de Harven, E. *et al.* (1977). Alternatives to critical point drying. *IITRI/SEM* **1**:519.

Hesse, I. and W. Hesse (1978). Artifacts on the surface of articular cartilage due to drying methods. *Proc. 9th Int. Cong. EM* **3**:680.

Klainer, A. S. *et al.* (1974). Evaluation and comparison of techniques for examination of bacteria by SEM. *IITRI/SEM* p. 314.

Lamb, J. C. IV and P. Ingram (1979). Drying of biological specimens for SEM directly from ethanol. *SEM, Inc.* **3**:459.

Liepins, A. and E. de Harven (1978). A rapid method for cell drying for scanning electron microscopy. *SEM, Inc.* **2**:37.

Marchant, H. J. (1974). Scanning electron microscopy of algal cells. *IITRI/SEM* p. 351.

Maugel, T. K. *et al.* (1980). Specimen preparation techniques for aquatic organisms. *SEM, Inc.* **2**:57.

Nickerson, A. W. *et al.* (1974). Spores. In: *Principles and Techniques of Scanning Electron Microscopy*, Vol. 1, M. A. Hayat, ed. Van Nostrand Reinhold, New York, p. 159.

Parsons, E. *et al.* (1974). A comparative survey of techniques for preparing plant surfaces for the scanning electron microscope. *J. Micros.* **101**:59.

Polliack, A. *et al.* (1974). Scanning electron microscopy of human leukocytes. A comparison of air dried and critical point dried cells. *J. Med. Sci.* **10**:1075.

—— *et al.* (1973). Comparison of air drying and critical point drying procedures for the study of human blood cells by SEM. *IITRI/SEM* p. 529.

Postek, M. J. (1975). Techniques for the preparation of phytoplaktonic organisms for scanning electron microscopy. *Tex. Rep. Biol. Med.* **33**:361.

Thornthwaite, J. T. and R. C. Leif (1974). The plaque cytogram assay. I. Light and scanning electron microscopy of immunocompetent cells. *J. Immunol.* **113**:1897.

Waterman, R. E. (1980). Preparation of embryonic tissues for SEM. *SEM, Inc.* 2:21.

Watson, L. P. *et al.* (1980). Preparation of microbiological specimens for scanning electron microscopy. *SEM, Inc.* **2**:45.

8. Handling Free-Living Cells

"Free-living cells" is an inclusive term for microorganisms, protozoa, blood cells, *etc.*: All of these cells have small dimensions. In comparison to a bulk tissue specimen, which is easily manipulated, free-living cells usually are microscopic, and thus require different handling methods. Although the techniques discussed below may appear trivial, they largely determine the success or failure of a study. These methods also tend to concentrate the sample to the point that a reasonable number of cells may be observed per specimen stub; clean the specimen and remove excess detritus; and, finally, leave the specimen on a smooth substrate that does not interfere with SEM observation. The techniques are micropipetting, centrifugation, and filtration. The choice of one technique is determined primarily by the size of the specimen. It is stressed that these methods are adjunct to fixation and drying techniques: For example, the specimen is filtered prior to chemical fixation. Reviews of specific cells will appear below; general published reports of manipulation of free-living cells are in Marchant (1973), Lung (1974), Nickerson *et al.* (1974) Albrecht and MacKenzie (1975), Baccetti (1975), de Harven *et al.* (1975), Hayes and Pawley (1975), and Sanders *et al.* (1975).

MACROSCOPIC ORGANISMS

Specimens just-resolvable by the unaided eye are in the 0.2–1-mm size range—*e.g.*, nematodes (Tombes *et al.*, 1978) and some foraminifera (LeFurgey, 1978). When the organism is suspended in liquid, it is easiest to transfer with a micropipette (1.5-mm inside diameter) from one solution to another. The passage must be made gently to avoid external damage and also to avoid carrying excess volume of solution. Alternatively, the sample is held in a nylon mesh or metal basket for transfer (For various specimen holders see, *e.g.*, Bronskill, 1970; Baker, 1972; Baker and Princen, 1975; Bassett and Pendergrass, 1975; Newell and Roath, 1975; Brown and Teetsov, 1976; Cohen, 1976; Sweney *et al.*, 1976; and Hayunga, 1977). These methods ensure that the specimen will not be lost during transfer or critical-point drying.

Frequently, suspended macroscopic specimens and microscopic free-living cells are coated with extraneous material (*e.g.*, detritus, mucous, *etc.*) which will interfere with a surface analysis if it is not removed. When additional suspending medium is available, *e.g.*, seawater, it should be filtered through a 0.2-μm-pore membrane filter and used as the rinsing medium. Maugel *et al.* (1980) list numerous washing media and the types of samples they may be used with; they also note that some extraneous material may be removed during fixation or dehydration, rather than prior to fixation. In any situation it is essential that the washing not be disruptive. Maugel *et al.* (1980) also describe relaxation methods and anaesthetics for larger microorganisms that, otherwise, may distort during fixation. Following washing, the cells are conventionally fixed, osmicated, dehydrated, and dried. Various preparation methods for aquatic organisms are outlined by Marchant (1973 and 1974), Maugel *et al.* (1980), Pedersen *et al.* (1980), and Small and Maugel (1978).

MICROSCOPIC CELLS AND ORGANISMS

Smaller specimens, resolvable with the light microscope and in the tenth-of-a-millimeter size range and smaller, include blood cells and bacteria. In this range, it is desirable to concentrate the number of cells for a representative sampling and to ensure that discrete particles may be observed. Either centrifugation or filtration achieve these goals.

Cells in the tenth-of-a-millimeter range have sufficient mass to settle to the bottom of a specimen holder without centrifugation, whereas smaller organisms, *e.g.*, bacteria, require centrifugation for concentration. Mild forces may be used to concentrate a specimen during the prefixation cleaning—*i.e.*, the cells are centrifuged and the supernatant decanted, resuspended in washing medium, centrifuged, *etc.* This is significantly more rapid than waiting for cells to naturally settle. Albrecht *et al.* (1978) purified, separated, and cleaned different blood-cell types by centrifugation, and then permitted each fraction to settle onto a coverslip. (The latter is a much smoother specimen support than a stub.) Various other modifications employing centrifugation of blood cells have been used (see Bessis and Weed, 1972; Dewar *et al.*, 1976; Newell *et al.*, 1978; and Carr and Toner, 1979).

Following cell cleaning by resuspension and centrifugation, the resultant pellet may be fixed, dehydrated, dried, and then dusted over a polished stub or glass coverslip. This method is tedious, however, because the specimen usually is resuspended at each preparation step, then recentrifuged to avoid sample loss. It is also difficult to obtain an even distribution of the dried specimen across a substrate, in that individual particles tend to clump together. Ideally, an even distribution of discrete cells on an amorphous substrate is preferred. Following separation of the desired specimen phase and cleaning by centrifugation, the suspended sample is filtered onto a membrane filter (*e.g.*, Newell *et al.*, 1978). This will both concentrate the sample to the appropriate loading (not too much to avoid particle overlap, but sufficient for the sampling to be representative of the bulk), and provide a featureless substrate for SEM observation—*i.e.*, unless hand polished, most specimen stubs are too rough. Many researchers have used filters as supports for a wide variety of specimens: For example, Munawar and Bistricki (1979) filtered nanoplankton crytonomads, Newell *et al.* (1978) worked with lymphocytes; and numerous authors have filtered microbes (*e.g.*, Bibel and Lawson, 1972a,b; Kormendy and Wayman, 1972; Todd and Kerr, 1972; Klainer *et al.*, 1974; Kormendy, 1975; Sanders *et al.*, 1975; McKee, 1977; Drier and Thruston, 1978; Gallagher and Rhoades, 1979; Watson *et al.*, 1980). As will be discussed, the filter is simply transferred from one processing solution to the next, much like bulk specimens.

Although a wide variety of filter types are available, only a few are amenable to SEM examination. Fibrous mats, such as common filter paper, cellulose-fiber, or glass-fiber filters are unsuitable because the specimen enters the filter matrix, rather than remaining on the surface. Porous filters are a slight improvement over fibrous mats, in that these are labyrinths of connecting passages having a specific diameter (variable from 0.2 to 8.0 μm). Examples are cellulose membrane filters (*e.g.*, trade name Millipore) and silver membranes. If the pore size is significantly smaller than the relevant specimen size, most of the particles will remain at the filter surface rather than enter the matrix. However, these filters are rough-surfaced when observed in the SEM and detract from the quality of the image.

The most suitable filter for SEM specimens, especially those in the 0.45 μm and smaller pore size range, are nuclear membranes (*e.g.*, trade name Nucleopore). These are smooth-surfaced polycarbonate membranes characteristically having linear, uniform, predominantly unconnected pores of diameter 0.2–8 μm. Porter (1977) discusses Nucleopore filters and their various applications, while Todd and Kerr (1972) and Watson *et al.* (1980) photographically illustrate different filter types. These membrane filters are more amenable for SEM because the smooth surface is interrupted only by the pore openings—meaning that the specimens lie only on the filter surface, and particles are not entrapped by the filter matrix (provided, of course, that the mean pore diameter is smaller than the specimen). Because the pore size is quite uniform, the pores may be used as rough indicators of dimension.

Another advantage of nuclear pore filters is that energy-dispersive x-ray analysis of supported particles is easy, because the filters's low molecular weight does not induce high noise levels. Thus, quick qualitative analyses are possible, but more exact x-ray analyses do require background subtractions. In any x-ray analysis, the filtered specimen must be mounted on a carbon planchet to exclude x-rays originating from the metal stub. That is, where small particles are mounted on a metal stub, the particle and the underlying stub both contribute to the x-ray signal: The effective excitation volume is the particle plus underlying surface.

Nuclear pore and membrane filters are readily adapted for both SEM and light microscope observations. Filters may be rendered invisible by mounting in an immersion fluid of characteristic refractive index for light microscopy: The refractive indices of Nucleopore filters are 1.584 and 1.614, while that of Millipore is 1.507. Thus, the same filtered specimen may be readily observed by light or electron optics. Special dissolution methods for TEM are also available (*e.g.*, see Chatfield and Dillon, 1978).

The major disadvantage of nuclear pore filters is their high expense, which is inversely proportional to pore size. Also, particle size will affect filtering efficiency; a high concentration of small particles decreases efficiency. The filters also tend to accumulate static charge, which may affect collection efficiency in that small particles are repelled. Because the filters are fragile (∿10-μm thickness), care must be taken to avoid mechanical damage. Finally, any filter to be examined by the SEM must be coated (sputter coatings are most common): Uncoated membranes and nuclear pore filters tend to "burn" during examination, causing charge accumulation and severe degradation of resolution. Even the coated filters tend toward residual halation, although it is not debilitating. In comparison, silver membranes are inherently conductive and may not require coating, especially if cell conductivity is enhanced by metallic impregnation (Bemrick and Hammer, 1978).

The general method for cleaning suspended specimens and concentrating them by filtration is as follows.

1. Clean the particle surface as discussed earlier, or with the Kormendy (1975) method. (Note: This method is used for removing debris, not mucus.):
   a. Centrifuge cells for 5 min at 1800 g (4000 rpm).
   b. Decant the supernatant and resuspend the pellet in 10 ml of the buffer to be used during fixation.
   c. Centrifuge for 5 min at 1800 g.
   d. Repeat steps b and c twice.
   e. Resuspend the pellet in fresh buffer or in glutaraldehyde fixative.
2. Filtration on Nucleopore membranes (Watson *et al.*, 1980):
   a. 13-mm-diameter, 0.4-μm-pore-size filters are labeled on one side with a pencil, submerged in water to eliminate static charge, and mounted in a Nucleopore filter holder.
   b. The sample suspension is drawn into a syringe and attached to the filter holder.
   c. Gently depress the plunger; cells will deposit over the filter surface.
3. Remove the filter, gently fold in half (specimen-side inward), wrap loosely in foil or place in a carrier as noted above, and continue processing through critical-point drying.

4. Gently cut the filter into the desired size and mount on stubs with double-sticky or conductive metal tape. Do not use liquid paints or adhesives: They will wet the filter.
5. Sputter coat the specimens and examine.

McKee (1977) and Watson *et al.* (1980) carried out primary and postfixation of the suspended particles before filtration; the filtration occurred during the postfixation water wash. They also used the syringe/filtration apparatus through dehydration and transition, but operator skill is essential in maintaining the specimen moist—otherwise, the cells will air-dry.

The primary fixation of suspended microbes is usually with 1–6% glutaraldehyde buffered with sodium cacodylate, phosphate, or *sym*-collidine, and postfixation with 1–2% osmium tetroxide (Williams *et al.*, 1973; Kormendy, 1975). Other fixation methods are summarized by Watson *et al.* (1980) and in Chapter 4: recall the usefulness of uranyl acetate en bloc staining for bacteria. Dehydration and critical-point drying follow conventional methods, as does freeze-drying. These methods are covered in their respective chapters.

REFERENCES

Albrecht, R. M. *et al.* (1978). Identification of monocytes, granulocytes, and lymphocytes: correlation of histological, histochemical, and functional properties with surface structures as viewed by scanning electron microscopy. *SEM, Inc.* 2:511.

—— and A. P. MacKenzie (1975). Cultured and free-living cells. In: *Principles and Techniques of Scanning Electron Microscopy* Vol. 3, M. A. Hayat, ed. Van Nostrand Reinhold, New York, p. 109.

Baccetti, B. (1975). Spermatozoa. In: *Principles and Techniques of Scanning Electron Microscopy*, Vol. 4, M. A. Hayat, ed. Van Nostrand Reinhold, New York, p. 94.

Baker, F. L. and L. H. Princen (1975). Positive displacement holder for critical point drying of small particle materials. *J. Micros.* **103**:393.

Baker, N. V. (1972). A microstrainer-syringe combination for handling specimens in the 1 mm^3 range in fluid changes during straining processes. *Stain Technol.* **47**:105.

Bassett, L. A. and R. E. Pendergrass (1975). A method for handling free cells through critical point drying. *J. Micros.* **109**:311.

Bemrick, W. J. and R. F. Hammer (1978). Scanning electron microscopy of damage to the cecal mucosa of chickens infected with *Eimeria tenella. Avian Dis.* **22**:86.

Bessis, M. and R. I. Weed (1972). Preparation of red blood cells (RBC) for SEM: A survey of various artifacts. *IITRI/SEM* p. 289.

Bibel, D. J. and J. W. Lawson (1972a). Scanning electron microscopy of L-phase streptococci. I. Development of techniques. *J. Micros.* **95**:453.

—— and J. W. Lawson (1972b). Scanning electron microscopy of L-phase streptococci. II. Growth in broth and upon millipore filters. *Can. J. Microbiol.* **18**:1179.

Bronskill, J. F. (1970). Fine-mesh stainless-steel gauze containers for fluid processing of very small free-floating specimens. *Stain Technol.* **45**:87.

Brown, J. A. and A. Teetsov (1976). Some techniques for handling particles in scanning electron microscope studies. *IITRI/SEM* **1**:385.

Carr, K. E. and P. G. Toner (1979). Scanning electron microscopy of macrophages: a bibliography. *SEM, Inc.* **3**:637.

Chatfield, E. J. and M. J. Dillon (1978). Some aspects of specimen preparation and limitations of precision in particulate analysis by SEM and TEM. *SEM, Inc.* **1**:487.

Cohen, W. D. (1976). Simple magnetic holder for critical point drying of microspecimen suspensions. *J. Micros.* **108**:221.

de Harven, E. *et al.* (1975). New observations on methods for preparing cell suspensions for scanning electron microscopy. *IITRI/SEM* p. 361.

Dewar, C. L. *et al.* (1976). A simple method of processing erythrocytes for SEM. *Am. J. Clin. Path.* **66**:760.

Drier, T. M. and E. L. Thurston (1978). Preparation of aquatic bacteria for enumeration by scanning electron microscopy. *SEM, Inc.* **2**:843.

Gallagher, J. E. and K. R. Rhoades (1979). Simplified preparation of mycoplasmas, an acholeplasma, and a spiroplasma for scanning electron microscopy. *J. Bact.* **137(2)**:972.

Hayes, T. L. and J. B. Pawley (1975). Very small biological specimens. In: *Principles and Techniques of Scanning Electron Microscopy*, Vol. 3, M. A. Hayat, ed. Van Nostrand Reinhold, New York, p. 45.

Hayunga, E. G. (1977). A specimen holder for dehydrating and processing very small tissue samples. *Trans. Amer. Micros. Soc.* **96**:156.

Klainer, A. S. *et al.* (1974). Evaluation and comparison of techniques for examination of bacteria by scanning electron microscopy. *IITRI/SEM* p. 313.

Kormendy, A. C. (1975). Microorganisms. In: *Principles and Techniques of Scanning Electron Microscopy*, Vol. 3, M. A. Hayat, ed. Van Nostrand Reinhold, New York, p. 82.

—— and M. Wayman (1972). Scanning electron microscopy of microorganisms. *Micron* **3**:33.

LeFurgey, A. (1978). Scanning electron microscopic characterization of recent arenaceous foraminifera. *SEM, Inc.* **2**:579.

Lung, B. (1974). The preparation of small particulate specimens by critical point drying: applications for scanning electron microscopy. *J. Micros.* **101**:77.

Marchant, H. J. (1973). Processing small delicate biological specimens for scanning electron microscopy. *J. Micros.* **97**:369.

—— (1974). Scanning electron microscopy of algal cells. *IITRI/SEM* p. 351.

Maugel, T. K. *et al.* (1980). Specimen preparation techniques for aquatic organisms. *SEM, Inc.* 2:57.

McKee, A. (1977). SEM in medical microbiology–an overview. *IITRI/SEM* 2:239.

Munawar, M. and T. Bistricki (1979). Scanning electron microscopy of some nanoplankton cryptomonads. *SEM, Inc.* 3:247.

Newell, D. G. *et al.* (1978). Surface morphology and lymphocyte maturation. *SEM, Inc.* 2:569.

—— and S. Roath (1975). A container for processing small volumes of cell suspensions for critical point drying. *J. Micros.* **104**:321.

Nickerson, A. W. *et al.* (1974). Spores. In: *Principles and Techniques of Scanning Electron Microscopy*, Vol. 1, p. 159. M. A. Hayat, ed. Van Nostrand Reinhold, New York.

Pedersen, M. *et al.* (1980). X-ray microanalysis of marine algae. *SEM, Inc.* 2:565.

Porter, M. C. (1977). A novel membrane filter for the laboratory. In: *Laboratory Instrumentation*, Series II, Vol. IV, Int. Sci. Communications, Inc., Fairfield, Connecticut, p. 286.

Sanders, S. K. *et al.* (1975). A high yield technique for preparing cells fixed in suspension for scanning electron microscopy. *J. Cell Biol.* **67**:476.

Small, E. B. and T. K. Maugel (1978). Observations on the permanence of protozoan preparations for scanning electron microscopy. *IITRI/SEM* 2:123.

Sweney, L. R. *et al.* (1976). Some methods for more efficient processing of SEM specimens. *J. Micros.* **108**:335.

Todd, R. L. and T. J. Kerr (1972). Scanning electron microscopy of microbial cells on membrane filters. *Appl. Microbiol.* **23**:1160.

Tombes, A. S. *et al.* (1978). Extraction of nematodes from drinking water and the comparative detection efficiency of optical and scanning electron microscopy. *SEM, Inc.* 2:297.

Watson, L. P. *et al.* (1980). Preparation of microbiological specimens for scanning electron microscopy. *SEM, Inc.* 2:45.

Williams, S. T. *et al.* (1973). Preparation of microbes for scanning electron microscopy. *IITRI/SEM* p. 735.

9. Conductive Thin Films

INTRODUCTION

Biological specimens inherently have very low atomic weight composition and high resistance, meaning that during electron irradiation untreated specimens absorb the beam, releasing only very low-resolution information (Echlin, 1974 and 1978; Munger, 1977). This problem is overcome by making either the sample surface or entire sample conductive–*i.e.*, increasing its electron density sufficiently for data signals (secondary and backscattered electrons) to be released. Surface density and conductivity is increased by depositing a very thin metal film (*e.g.*, 250 Å of gold) by sputter coating or evaporating a heavy metal film over the specimen. Alternately, the density of the entire specimen is enhanced during preparation by treatment with mordants and heavy metal salts; *e.g.*, after post-fixation with osmium, the specimen is exposed to thiocarbohydrazide and osmium twice in series. The latter method is most useful when a metal film may interfere with surface structure, or for intramicroscope dissection. Metallic impregnation will be discussed in another chapter.

EVAPORATED THIN FILMS

Evaporated thin films are prepared under high vacuum by passing a current through a metal wire; as the temperature of the metal increases, thermal energy displaces atoms from the source, and these travel and deposit on anything in their path. This method of evaporation is referred to as "resistance heating." The metal atoms will then form an even, continuous film on the specimen surface: To ensure film continuity, the specimens are held on an eucentrically rotating specimen stage. In the literature, films accreted by this method are referred to as "rotary evaporation," "high vacuum evaporation," "thermal vacuum evaporation," and "bell jar evaporation" films. This is a modification of shadow casting, originally introduced by Williams and Wyckoff (1946) for enhancing the contrast of electron-transparent particles in the TEM.

METHOD OF EVAPORATION

The apparatus used for evaporation is the high-vacuum bell jar system (Figure 9-1). Essentially it is a large glass jar evacuated by a rotary and diffusion pump in series, arranged in exactly the same manner as in the SEM vacuum system. The baseplate of the system has air-tight ports for entrance of high-voltage cables, through which a current can be passed, to a conductive holder of the metal to be evaporated (Shiflett, 1968). The holder is usually tungsten wire shaped into a basket for the evaporation of nonrefractory metals. Tungsten baskets are commercially available, but they are easily hand-made, using 0.5–1.0-mil diameter pure tungsten wire (~10 cm long) and a wood screw. Position the screw at the center of the wire, then wrap the wire around the screw threads. If the wire is brittle,

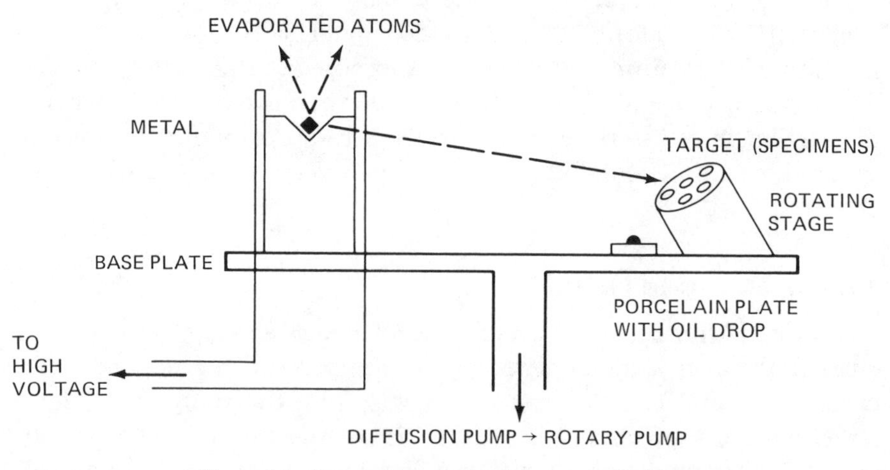

Figure 9-1. Evaporative coating in a vacuum bell jar.

heat it in a Bunsen burner to make it more malleable (Bradley, 1967). Baskets become brittle with use—but they may be used repeatedly, provided the wire is continuous. As will be discussed, nonmetallic carbon films are employed for x-ray elemental analysis of biological materials; in this situation, spring-held carbon electrodes eliminate the need for a substrate. (The two electrodes are simultaneously the substrate and evaporant.)

The specimens are located at least 10 cm from the source to avoid heat damage (Zingsheim *et al.*, 1970; Rowsowski and Glider, 1977; and Slayter, 1980). Temperatures as high as 1800°C are reached in nonrefractory metal evaporation—which obviously could damage fragile biological material, although Braten (1978) noted that the thermal energy of evaporated atoms is lower than that of sputtered atoms. (Temperature was monitored by positioning a thermocouple on the specimen stage; Clark *et al.*, 1976.) To avoid shadowing effects and to maintain equal thickness of the film across the specimen surface, the samples are mounted on a small platform which randomly rotates and tilts. Figure 9-2 shows the effect of this method; if the specimen were immobile during evaporation, the film would be noncontinuous and uneven in thickness.

A 10–15-cm length of the metal to be evaporated (*e.g.*, a 7-mil diameter platinum wire) is coiled and placed in the tungsten basket. The specimens, affixed to stubs, are placed in the rotary stage. Adjacent to the specimens, place a clean porcelain plate with a drop of vacuum oil or grease on it (Figure 9-1); this is a crude gauge for estimating film thickness. As evaporation proceeds the area beneath the oil drop remains white while the surrounding area becomes dark with metal. A light gray color roughly corresponds to a film thickness of ~100 Å. More exact methods for measuring thin films are in "Film Thickness." Evacuate to a minimum of 10^{-4} torr: As a rule,

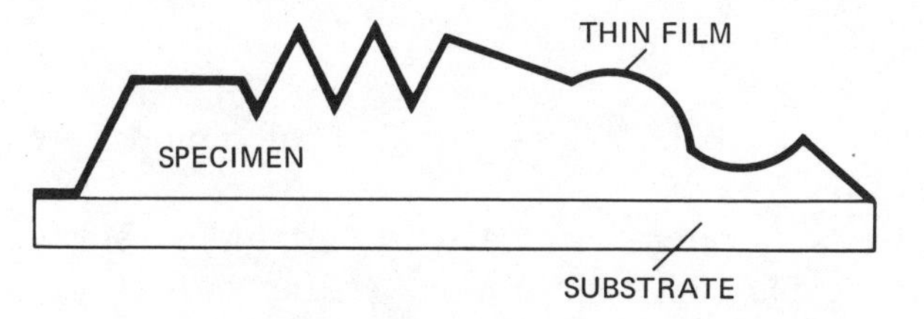

Figure 9-2. Thin films reproduce surface morphology.

the higher the vacuum, the finer the film. The following information on the formation of thin films resulting from resistance heating is compiled from a number of sources; in general, reference to any text of thin-film technology provides this data (*e.g.*, Shiflett, 1968; Holland, 1970; Neugebauer, 1970; Hayek and Schwabe, 1971 and 1974; Abermann *et al.*, 1972; Nagatani and Saito, 1974).

When a progressively higher current is passed through the tungsten basket, the temperature of the contained metal wire will rise; at a point just below melting, metal atoms will vaporize and follow a straight-line trajectory away from the source. The pathway of the vaporized atoms is linear because of the high-vacuum condition, and they contain sufficient thermal/kinetic energy to travel until they are intercepted by anything within the bell jar. The metal atoms continue to migrate until they are accommodated by a surface (*e.g.*, the specimen), with one of the following modes operational:

1. The excess thermal energy is transferred to the considerably cooler specimen surface or bell jar interior; surface mobility may thus be further reduced by cooling the specimen (Slayter, 1976). Cooling forms smaller and more uniformly sized clusters per unit area.
2. The vaporized metal atoms return to the ground state by entrapment at binding sites on the specimen. Henderson and Griffiths (1972) note that high-melting-point metals have strong internal binding characteristics and, thus, return very rapidly to the ground state.
3. As the number of free, mobile atoms increases, so does the probability that these atoms contact one another. These very small, isolated, atom clusters, referred to as "nuclei" in thin-film terminology, continue to grow as more metal is evaporated; and, over time, groups of clusters become continuous and form an intact thin film (Neugebauer, 1970).

In practice, it is probable that all three of these modes are functional during thin-film growth.

A completely different approach to suppressing surface mobility and increasing the film's amorphous character (*i.e.*, decreasing the size of crystallites, which increases ultimate fineness and resolution), is to simultaneously evaporate a relatively high-melting-point non-refractory metal, platinum (m.p. 1755°C), with carbon (m.p. 3800°C). This method was introduced by Bradley (1958 and 1959), who theorized that the carbon atoms interfere with the growth of large platinum crystallites; thus, a much finer-grain film is obtained. It was difficult in the past to routinely predict high degrees of success with this method, because one could not control the ratio of carbon to platinum evaporated. Harris (1975) devised an apparatus adaptable to any bell jar which reliably maintains the ratio of each element during evaporation. As will be discussed, mixtures of metals (*e.g.*, Au/Pd) serve the same purpose.

A number of parameters of thin films, whether they are grown by evaporation or any other method, become apparent. First, because the purpose of the thin film is to increase the conductivity of the specimen surface, the film must be continuous (filling in gaps or holes in the specimen), but the thickness must not be large enough to obscure surface detail; that is, one aims for a thin, but uninterrupted film. If isolated clusters existed, conductivity and consequently resolution would be poor; too thick a film will obscure surface features and, again, degrade resolution (Abermann *et al.*, 1972; Abermann and Salpeter, 1974). The minimum thickness of a film is characteristic of the evaporative metal itself: The degree of granularity is inversely proportional to the melting point–*i.e.*, the higher the metal's melting point, the finer the film. For example, gold, with a melting point of 1063°C, will form a grainy film minimally 50–75 Å thick, while platinum (m.p. 1755°C) forms small clusters and more featureless films. Intermediate film thicknesses are obtainable with other nonrefractory metals, such as palladium (melting point 1550°C, ~50 Å thick); a method for suppressing granularity is to evaporate alloyed gold/palladium, gold/platinum, platinum/palladium (Molcik, 1967), or as mentioned earlier, platinum/carbon. Pure gold evaporated films are commonly used in SEM; in most situations, it is unnecessary to employ very fine films unless high resolution (~50 Å) is necessary. In the latter case, platinum films are used (*e.g.*, see Broers and Spiller, 1980).

Even finer films may be prepared by refractory evaporation (Slayter, 1976 and 1980). The refractory metals used in electron microscopy are tungsten (Slayter, 1976), tungsten trioxide (Shved and Cylo,

1978) and tantalum/tungsten (Abermann and Salpeter, 1974; and Vasilier and Koteliansky, 1979). For the most part, these refractory metals have been used for TEM, although applications in very high-resolution SEM are becoming apparent (Slayter, 1980). Refractory thin films may be prepared by resistence heating (Hart, 1963; Hintermann and Begnin, 1967), sputtering (Kanaya and Hojou, 1974), or by evaporation from an electron gun (Abermann *et al.*, 1972; Hagler *et al.*, 1977; Slayter, 1978).

The metal films discussed, above, will degrade the detection of characteristic x-rays and cathodoluminescence–*i.e.*, low-energy data will be absorbed by the film. To increase conductivity without a loss of resolution of these signals, low-*Z* elements are used for thin films. Carbon evaporated from an electric arc under low-vacuum conditions permits such an analysis–but carbon has such a low atomic weight (12.01) that electron emission and, therefore resolution, is poor. Alternately, sputtered or evaporated aluminum films (atomic weight 26.98) are sufficiently conductive for good surface analysis, and are of low enough density to permit EDXRA (Na, Mg, Si, and Br cannot be detected). Aluminum films are prepared according to standard sputtering technique.

Carbon films are prepared in a low-vacuum ($\sim 10^{-2}$ torr) bell jar apparatus, using two spectroscopically pure electrodes–one pointed and the other squared off, or both pointed; and one electrode (pointed) spring-held against the other: The degree of compression will determine the film thickness. The specimens are mounted on

Table 9-1. Elements Used for Evaporation.

ELEMENT	AT. NO.	DENSITY (kg/m^3 × 10^{-3})	M.P. (°C)	B.P. (°C)	GRANULATION
Aluminum	13	2.70	...	...	...
Carbon	6	2.26	3800	2681	...
Gold	79	19.3	1063	1465	Coarse
Gold/Palladium	...	16.1	...	~1566	Coarse
Palladium	46	12.0	1550	~1566	Coarse
Platinum	78	21.5	1755	2090	Very fine
Platinum/Carbon	...	...	...	...	Very fine
Platinum/Palladium	...	19.4	...	~2090	Fine
Tungsten	74	...	...	...	...
Tungsten/Tantalum	...	19.3	...	3582	Very fine

the conventional tilt-and-rotate stage, the electrodes positioned, and high vacuum attained. Inert argon gas is bled into the chamber until the vacuum is $\sim 10^{-2}$ torr; it may be necessary to flush the chamber several times with argon to ensure a pure atmosphere. Evaporate at 30 V and 20 amp; a more continuous film will result if the evaporation is pulsed, *i.e.*, turned on and off. Characteristics of the more common materials used for evaporation are summarized in Table 9-1; much of this data was compiled from Echlin (1978).

Film Thickness

The "thickness" of an evaporated film is considered either as average thickness, defined as the real film thickness measured by quartz monitors, or as the mass thickness, which refers to the scattering potential of the film. Most thin-film researchers refer to "mass thickness," which takes into account variations within ultrathin film structure (*e.g.*, Slayter, 1980). Regardless of which terminology is used, film thickness is very important because it sets a limit to resolution—assuming that specimen preparation and the SEM are "perfect" (*i.e.*, that preservation is optimal and resolving performance is highest, respectively). For example, if one requires 60-Å point-to-point resolution, a thin film must be $\leqslant$60 Å thick: This simplification ignores penetration and excitation depths, as well as other factors, but applies to a perfect system. Routinely, the constraint is not a limiting factor in nonresearch situations.

Methods for measuring film thickness are nontrivial. Flood (1980) has an excellent review of these methods, and classifies the methods as before, during, or after evaporation, and *in situ* measurements. In the first situation, calculations are performed with the following factors (Flood, 1980):

$$T = \frac{W}{4R^2 \cdot d}$$

where

T = film thickness
W = amount of evaporated source material
R = source–substrate distance
d = density of deposited material

Unfortunately, this method does not take into account other parameters—*e.g.*, whether the evaporative metal completely evaporates, or alloys with the filament; and assumes that the target is perpendicular to the source (Echlin, 1978). As will be discussed, however, a dependable calculation exists for predicting sputtered film thickness.

The tools necessary for measuring thickness during evaporation are either a vacuum microbalance (Pearson and Wadsworth, 1965), or quartz crystal oscillators having a resonant frequency inversely proportional to film thickness (Chopra, 1969; Glang, 1970). Both of these methods are commonly encountered in thin film physics laboratories, but are fairly uncommon in SEM laboratories (Peters, 1980), primarily because neither method is effective for discontinuous films.

Thickness measurement, after deposition, involves either direct measurement of the film on the specimen by x-ray microanalysis, or deposition on a flat test substrate held near the specimen within the apparatus. With microanalysis, film thickness is proportional to emitted x-ray intensity (Priyokumos-Singh *et. al.*, 1976): The specimen must be flat where the spectrum is recorded. Hohn (1977) and Niedrig (1978) employed backscattered electrons in much the same manner.

Flat substrates with attached films may be cross sectioned and examined by TEM (Brunner *et. al.*, 1975). Inaccuracies develop, however, from mechincal deformation induced by thin sectioning. Alternatively, a thin film is deposited on latex spheres and these are examined by TEM (Schur *et al.*, 1967; Roli and Flood, 1978). Flood (1980) contains an excellent discussion of this methodology.

Very accurate thickness measurements may be conducted on flat substrates using two- or multiple-beam interference methods (*e.g.*, Pliskin and Zanin, 1970; Flood, 1980). Interferometry is standard in situations where high accuracy is required—*e.g.*, in thin film physics.

Any of these methods will define the conditions necessary for reproducible film thicknesses within moderate accuracy limits. It must be stressed that the above methods are not routine, and typically apply only to research situations where very high-resolution microscopy is important. In more routine situations, a thicker film (*i.e.*, one thicker than the minimum required for continuity) is used. As the instrumental parameters controlling resolution are reduced, film thickness will become more important.

Good evaporation technique requires that a few general rules be followed. First, the source becomes extremely bright during evaporation: Never look directly at the source without protective welding goggles; or protect your eyes with cobalt glass. Second, the bell jar apparatus must be thoroughly cleaned to ensure good vacuum technique. After each use, clean all coated surfaces (bell jar interior, baseplate, *etc.*) with acetone and lint-free cloths, and re-evacuate to high vacuum. If the interior is not cleaned, a metal/vapor/metal sandwich will build up, and the vapor will continuously leak out—making vacuum conditions progressively worse. Consequently, very poor, rough thin films result.

Sputtered Films

Another method for depositing conductive thin films is by sputter coating. With this technology, metal atoms are eroded from a target material by energetic plasmas of argon. A transfer of momentum from the argon ions to the surface layer of the target, usually gold, is sufficient for gold atoms to be ejected. Because this event occurs under low vacuum—meaning that the travelling gold atoms are deflected by interaction with gas molecules—a very continuous thin film will grow. On the other hand, precautions against heat damage must be observed, because the interaction between the plasma and target produces more thermal energy than a 1:1 ratio of incident-gas-ion: released-gold-atom (Holland, 1976; Panayi *et. al.*, 1977; Braten, 1978; Echlin *et. al.*, 1980). Wehner and Anderson (1970) and Echlin (1978) cleverly describe this situation as analogous to three-dimensional billiards. These events occur when the sputtering target is made the cathode, and the specimens are the anode; the effective accelerating voltage (*i.e.*, the acceleration of sputtered atoms toward the specimen) is 2.3 keV. The negative potential of the sputtering material forces the release of electrons which ionize the contained gas, thereby giving rise to a purple discharge. If the potential difference, or high tension, is too high, these electrons will be attracted toward the specimen and damage its surface.

Relatively large ions of argon (atomic number 18) are used as the sputtering plasma, because large mass is necessary to displace the even-larger metal atoms. Nitrogen (atomic number 7) is sometimes used, but is the practical limit for efficiency. Sputtering with impure gas or air decreases efficiency, because water vapor and carbon dioxide decompose to oxygen molecules, diminishing the deposition rate. The deposition rate is also influenced by substrate temperature: Although efficiency increases with temperature, the probability of thermal damage proportionately increases. Echlin *et al.*, (1980) have very effectively demonstrated that cooling the specimen and increasing the duration of sputtering yields higher quality films (*i.e.*, they are smooth and relatively featureless). The sputter coater must also be free from contamination for efficient rates; for example, the specimen chamber is flushed with argon several times to ensure a pure atmosphere, and oil contamination from the rotary pump must be avoided (de Harven *et. al.*, 1978).

A number of different sputtering methods and theories are discussed in detail by Wehner (1969), Maissel (1970), Wehner and Anderson (1970), and Sherman *et al.* (1973). The methodology discussed above is generally referred to as plasma sputtering, and employs either a diode, triode, or radio-frequency sputter apparatus. The diode, or direct-current (DC) sputter coater is the commercially available apparatus usually found in SEM laboratories (Figure 9-3). It consists of a small bell jar evacuated by a rotary pump, the target functioning as a cathode, and a cooled specimen holder functioning as the anode (Echlin, 1975; Echlin and Hyde, 1972). Panayi *et al.* (1977) designed a cool diode unit which significantly reduces heat damage by incorporating an electron deflector directed away from the specimen. They were successful in coating wax specimens (m.p. 32°C) which would melt under other sputtering conditions.

Triode sputter coaters employ a plasma formed separately from the specimen, which essentially becomes a second cathode. (Thus, the triode consists of the anode or target, and two cathodes: one the specimen at ground and the second a separate electrode forming the plasma.) Again, heat damage is reduced, and because the system is operational at better than 10^{-2} torr, contamination is lower (Ingram *et al.*, 1976).

Another method of sputtering is by radio-frequency. An alternating voltage, modulated with a 10-MHz frequency, is applied over the specimen; this induces radio-frequency on the specimen surface. Because this method is used for sputtering nonconductive materials

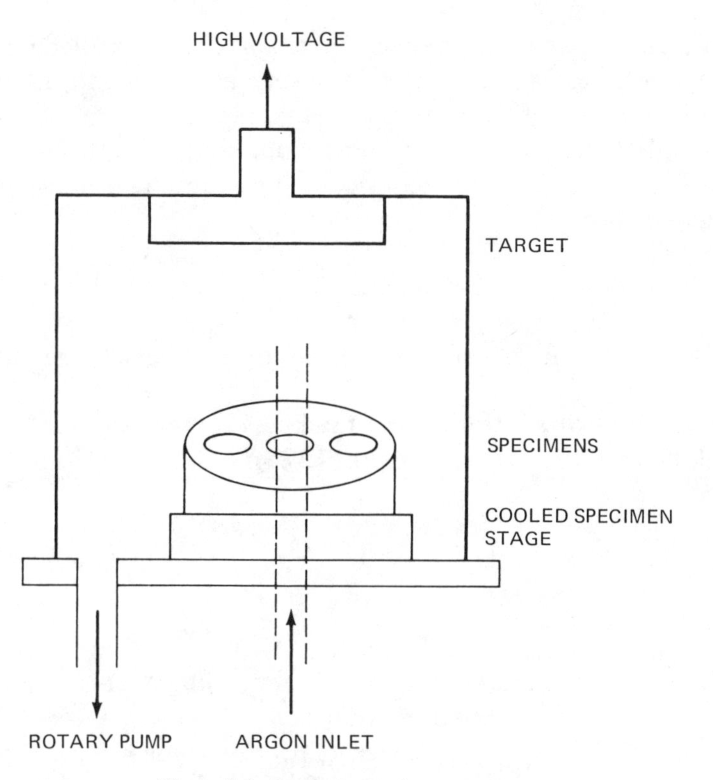

Figure 9-3. Diode sputtering apparatus.

(*e.g.*, cleaning of specimen surfaces), it has no real advantage for SEM.

Ion-beam sputtering involves oblique deposition of the sputtered metal (again using argon) under high vacuum (Geller *et. al.*, 1979; Franks *et. al.*, 1980). It employs a cold cathode discharge, eliminating the problems associated with electron bombardments, and thus heat damage. This instrument is more commonly found in materials science than in biological laboratories.

The general method for plasma sputtering first involves evacuating the specimen chamber to the 10^{-2} torr range and pumping out of residual vapors. Because adhesives holding the specimen to the stub (*e.g.*, double-stick tape or glue), and possibly the specimen itself, contain volatile vapors, it may take longer than expected to attain low vacuum. Second, the specimen chamber is backfilled with argon and reevacuated several times to ensure a pure argon atmosphere: When volatile samples are to be coated, repeated flushings are necessary, while "clean" specimens may require only two flushings. While this is occurring, the specimen is cooled by running water. The actual sputtering then follows this sequence:

1. Pump down to $\sim 10^{-2}$ torr.
2. Turn on the high voltage to $\sim$2.5 keV.
3. Admit argon to the chamber to a plasma discharge current of $\sim$20 mamp.
4. Duration of sputtering and thickness is predetermined using the following calculation (Echlin, 1978):

$$T = \text{mA} \times \text{keV} \times t \times \text{k}$$

where

T = thickness (Å)
mA = plasma current (milliamperes)
t = time (min)
k = gas constant. For argon k = 5.

5. After sputtering, turn off the rotary pump/sputter coater, backfill with argon to atmospheric pressure, and examine the specimens.

Gold or palladium mixtures are commonly used as the sputtering metal, although other metals are also available. The thin films grow in much the same manner as evaporated films with the following two exceptions: Individual or clustered metal atoms may slightly penetrate the specimen surface and, because of low-vacuum conditions, the metal atoms collide with the argon gas, resulting in a very continuous coating of rough surfaces. (The specimen stage is immobile.)

The characteristic of sputtered films forming a continuous coating on specimen surfaces is very much an advantage for biological materials. Because so many biological surfaces are nonconductive and rough (*e.g.*, microvilli of intestine and surface features of diatoms), sputtered films effectively render them conductive. (Slobodrian *et al.*, 1978 include a method for accurately sputtering specimens with cavities.) The desired thickness is simply calculated and the sputtering apparatus preset for desired conditions.

On the other hand, artifacts become apparent if this technique is abused. Thermal damage will result if the sputtering rate is too high, especially when older instruments that do not have a cooled specimen stage are used (De Nee and Walker, 1975; Echlin, 1975; Holland, 1976; Rowsowski and Glider, 1977). By decreasing the sputtering rate and increasing the duration, good coatings are obtainable (Panayi *et. al.*, 1977). Second, contamination artifacts may interfere with the specimens inherent surface features; this can be traced to inadequate argon flushing (the specimen or adhesive out-gases) or backflow of oil from the rotary pump into the chamber (Holland, 1976; de Harven *et. al.*, 1978). If the apparatus is not equipped with an oil trap between the pump and specimen chamber, the chamber must be rapidly backfilled to avoid oil backstreaming (Simmens, 1975). Likewise, the specimen chamber must be cleaned with lint-free cloths and acetone to avoid contamination.

COMPARISON OF EVAPORATION AND SPUTTERING

There is a great deal of controversy when sputtered and evaporated coating, and their effects on surface morphology, are compared (*e.g.*, De Nee and Walker, 1975; Ingram *et al.*, 1976; Munger, 1977; Rowsowski and Glider, 1977; Echlin, 1978). If either method is abused, artifacts will result: however, with proper attention–for example, to ambient temperature–artifacts may be interpreted. In general, cool sputtering with gold or gold-palladium is recommended for rendering any biological material conductive for surface analysis, whereas carbon evaporation is recommended for specimens to be elementally analyzed. It may be necessary to evaporate metals when very high SEM resolution ($\leqslant$50-Å) is required, but most SEMs guarantee resolutions of 60–80 Å. As instrumetation achieves greater resolution, *e.g.*, when brighter, more efficient sources are routinely used (LaB_6 and field-emission guns), it may be necessary to re-evaluate metal coatings. In this situation, it is most likely that smoother films, such as are obtained by evaporation, will become more important. A completely different approach, although not universally applicable, is to render the specimen conductive by reaction with heavy metals; this topic is discussed in Chapter 10.

REFERENCES

Abermann, R. and M. M. Salpeter (1974). Visualization of desoxyribonucleic acid molecules by protein film adsorption and tantalum-tungsten shadowing, *J. Histochem. Cytochem.* **22**:845.

—— *et al.* (1972). High resolution shadowing. In: *Principles and Techniques of Electron Microscopy*, Vol. 2, M. A. Hayat, ed. Van Nostrand Reinhold, New York, p. 197.

Bradley, D. E. (1958). Simultaneous evaporation of platinum and carbon for possible use in high resolution shadow casting for electron microscopy. *Nature* **181**:875.

—— (1959). High-resolution shadow casting techniques for the electron microscope using the simultaneous evaporation of platinum and carbon. *Brit. J. Appl. Phys.* **10**:198.

—— (1967). Replica and shadowing techniques. In: *Techniques for Electron Microscopy*, 2nd ed., D. H. Kay, ed. Blackwell, Oxford, p. 96.

Braten, T. (1978). High resolution scanning electron microscopy in biology: artefacts caused by the nature and mode of application of the coating material. *J. Micros.* **113**:53.

Broers, A. N. and E. Spiller (1980). A comparison of high resolution scanning electron micrographs of metal film coatings with soft x-ray interference measurements of the film roughness. *SEM, Inc.* **1**:201.

Brunner, M E. G. *et al.* (1975). A technique for sequential examination of specific areas of large tissue blocks using SEM, LM, and TEM. *IITRI/SEM* p. 333.

Chopra, K. L. (1969). *Thin Film Phenomena.* McGraw Hill, New York.

Clark, J. *et al.* (1976). Thin film thermocouples for use in scanning electron microscopy. *IITRI/SEM* **1**:83.

de Harven, E. *et al.* (1978). Sputter coating in oil-contamination-free vacuum for scanning electron microscopy. *IITRI/SEM* **1**:167.

De Nee, P. B. and E. R. Walker (1975). Specimen coating technique for the SEM–a comparative study. *IITRI/SEM*, p. 225.

Echlin, P. (1974). Coating techniques for scanning electron microscopy. *IITRI/SEM*, p. 1019.

—— (1975). Sputter coating techniques for scanning electron microscopy. *IITRI/SEM*, p. 217.

—— (1978). Coating techniques for scanning electron microscopy and x-ray microanalysis. *SEM, Inc.* **1**:109.

—— *et al.* (1980). Improved resolution of sputter-coated metal films. *SEM, Inc.* **1**:163.

—— and P. J. W. Hyde (1972). The rationale and mode of application of thin films to non-conducting materials. *IITRI/SEM* p. 137.

Flood, P. R. (1980). Thin film thickness measurement. *SEM, Inc.* **1**:183.

Franks, J. *et al.* (1980). Ion beam thin film deposition. *SEM, Inc.* **1**:155.

Geller, J. D. *et al.* (1979). Coating by ion sputtering deposition for ultrahigh resolution SEM. *SEM, Inc.* **2**:355.

Glang, R. (1970). Vacuum evaporation. In: *Handbook of Thin Film Technology*, C-7. L. I. Maissel and R. Glang, eds. McGraw Hill, New York.

Hagler, H. F. *et al.* (1977). A simple electron beam gun for platinum evaporation. *J. Micros.* **110**:149.

Harris, W. J. (1975). A universal metal and carbon evaporation accessory for electron microscope techniques and a method for obtaining repeatable evaporation of platinum–carbon. *J. Micros.* **105**:265.

Hart, R. G. (1963). A method for shadowing electron microscope specimens with tungsten. *J. Appl. Phys.* **34**:434.

Hayek, K. and U. Schwabe (1971). Die anwendung einer hochanflusenden beschattungsmethode zur urtersuching der schichtbildung von Au auf NaCl bei tiefin temperaturen. *Z. Naturforsch.* **26a**:1879.

—— and U. Schwabe (1974). Applicating high resolution shadow casting to the study of nucleation and growth of gold on sodium chloride. *J. Vac. Sci. Technol.* **9**:507.

Henderson, W. J. and K. Griffiths (1972). Shadow casting and replication. In: *Principles and Techniques of Electron Microscopy*, Vol. 2, M. A. Hayat, ed. Van Nostrand Reinhold, New York, p. 151.

Hintermann, H. E. and J. Begnin (1967). Tungsten shadow casting for electron microscopical specimens. *J. Sci. Inst.* **44**:207.

Hohn, F. T. (1977). Angular dependence of electron intensities backscattered by carbon films. *Optik* **47**:491.

Holland, L. (1970). *Vacuum Deposition of Thin Films.* Chapman and Hall, Ltd., London.

Holland, V. F. (1976). Some artifacts associated with sputter coated samples observed at high magnifications in the scanning electron microscope. *IITRI/SEM* **1**:71.

Ingram, P. *et al.* (1976). Some comparisons of the techniques of sputter (coating) and evaporative coating for scanning electron microscopy. *IITRI/SEM* **1**:75.

Kanaya, K. and K. Hojou (1974). Ion bombardment of suitable targets for atomic shadowing for high resolution electron microscopy. *Micron* **5**:89.

Maissel, L. I. (1970). Application of sputtering to the deposition of films. In: *Handbook of Thin Film Technology*, C-4. L. I. Maissel and R. Glang, eds. McGraw Hill, New York.

Molcik, M. (1967). Technique to reduce the grain size of shadowing alloys used in electron microscopy. *Praktical. Metallogr.* **4**:628.

Munger, B. L. (1977). The problem of specimen conductivity in electron microscopy. *IITRI/SEM* **1**:481.

Nagatani, T. and M. Saito (1974). Structure analysis of evaporated films by means of TEM and SEM. *IITRI/SEM* p. 51.

Neugebauer, C. A. (1970). Condensation, nucleation, and growth of thin films. In: *Handbook of Thin Film Technology*, C-8. L. I. Maissel and R. Glang, eds. McGraw Hill, New York.

Niedrig, H. (1978). Backscattered electrons as a tool for film thickness determination. *SEM, Inc.* **1**:841.

Panayi, D. N. *et al.* (1977). A cool sputtering system for coating heat-sensitive materials. *IITRI/SEM* **1**:463.

Pearson, S. and N. J. Wadsworth (1965). A robust torsion balance which can detect a force of 2×10^{-8} dyne. *J. Sci. Instr.* **42**:150.

Peters, K. R. (1980). Penning sputtering of ultrathin metal films for high resolution electron microscopy. *SEM, Inc.* **1**:143.

Pliskin, W. A. and S. J. Zanin (1970). Film thickness and composition. In: *Handbook of Thin Film Technology*, C-11. L. I. Maissel and R. Glang, eds. McGraw Hill, New York.

Priyokumas-Singh, S. *et al.* (1979). Thickness measurements of single and composite thin metal films using the X-ray fluorescence technique. *Thin Solid Films* **59**:51.

Roli, J. and P. R. Flood (1978). A simple method for the determination of thickness and grain size of deposited films as used in non-conductive specimens for SEM. *J. Micros.* **112**:359.

Rowsowski, J. R. and W. V. Glider (1977). Comparative effects of metal coating by sputtering and vacuum evaporation on delicate features of euglenoid flagellates. *IITRI/SEM* **1**:471.

Schur, K. *et al.* (1967). Auflösungsvermögen und Kontrast von Overflächenstufen bei der Abbildung mit einem Raster-Elektronenmikroskop (Stereoscan). *Z. Angew. Phys.* **23**:405.

Sherman, D. M., J. S. May and T. E. Hutchinson (1973). Observation of sputtered film growth. *J. Vac. Sci. Technol.* **10**:155.

Shiflett, C. C. (1968). Evaporated film. In: *Thin Film Technology*, R. W. Berry, P. M. Hall and M. T. Harris, eds. Van Nostrand, New York, p. 113.

Shved, A. D. and V. V. Cylo (1978). Use of tungsten trioxide for shadowing nucleic acid molecules. *Tsitol. Genet.* **12**:70.

Simmens, S. C. (1975). An observation on the metallizing of specimens for scanning electron microscopy using cathode sputtering. *J. Micros.* **105**:233.

Slayter, H. S. (1976). High resolution metal replicating of macromolecules. *Ultramicroscopy* **1**:341.

—— (1978). Fine features of glycoproteins by high resolution metal replication. In: *Principles and Techniques of Electron Microscopy*, Vol. 9, M. A. Hayat, ed. Van Nostrand Reinhold, New York, p. 175.

—— (1980). High resolution metal coatings of biopolymers. *SEM, Inc.* **1**:171.

Slobodrian, M. L. *et al.* (1978). Metallic deposition in specimens presenting cavities using the sputter coater. *J. Micros.* **112**:365.

Vasilier, V. D. and V. E. Kotelianski (1979). Freeze drying and high resolution shadowing in electron microscopy of *Escherichia coli* ribosomes. *Methods Enzymol.* **59**:612.

Wehner, G. K. (1969). Angular distribution of sputtered material. *J. Appl. Phys.* **31**:177.

—— and G. S. Anderson (1970). The nature of physical sputtering. In: *Handbook of Thin Film Technology*, C-3. L. I. Maissel and R. Glang, eds. McGraw Hill, New York.

Williams, R. C. and R. W. G. Wyckoff (1946). Applications of metallic shadow casting to microscopy. *J. Appl. Phys.* **17**:23.

Zingsheim, H. P. *et al.* (1970). Shadow casting and heat damage. *Proc. 7th Int. Cong. EM* **1**:411.

10. Uncoated Specimens

Various situations exist where coating of nonconductive specimens by evaporation or sputtering is undesirable. The need for good conductivity and, thus, resolution, has been previously discussed in other chapters (also see DeNee and Walker, 1975; Munger, 1977). Alternatives to metal films are the examination of uncoated frozen specimens on a cooling stage, without enhancing electron density; or metallic impregnation methods, in which conductivity is increased by treating specimens with additive metal salts.

Interest in the examination of uncoated specimens was motivated by the artifacts (*e.g.*, thermal damage) that may be incurred during evaporation or sputtering, and by the problem of generating a continuous film without obscuring surface features on rough samples. As discussed in Chapter 9, these artifacts are minimized if good technique is followed: Thus, the methods discussed below are not replacements for thin films; the vast majority of biological specimens analyzed morphologically are coated with thin films. In comparison, if one desired to examine morphology of the same specimen at different levels (*i.e.*, examine the surface, then dissect deeper into the tissue and re-examine), whole-specimen conductivity would be an advantage. Another example is if the thin film must be very thick to be continuous (because of a very rough surface; *e.g.*, fractured kidney specimens), surface details are hidden by the film. These alternative methods are discussed below, and Murphy (1978 and 1980) provides comprehensive reviews.

FRESH OR FROZEN SPECIMENS

Low-magnification studies of native specimens (*i.e.*, nonfrozen and untreated) are possible, provided that low accelerating voltages (<5 keV) are used. For example, the study of whole insects by SEM provides much greater depth of field than is possible with the light microscope. By examining the untreated specimen nondestructively (*i.e.*, as is), the entire insect may be examined in either instrument (Howden and Ling, 1974).

The general procedure for examining a fresh specimen is to first mount it with a conductive adhesive, thus providing contact to ground and reducing charge artifacts. Copper or aluminum tape is suitable, although the adhesive *per se* is nonconductive. A small amount of silver paint may be used, but the paint tends to creep by capillary action through the specimen. If it is possible to paint a strip of silver from the specimen to the stub in an area different from that of interest, very good contact to ground results. These manipulations are easily performed with a binocular microscope, forceps, and needles. If the specimen is very dry, it is useful to "hydrate" it by storage in a petri dish with a few sheets of damp filter paper. Panessa and Gennaro (1974) recommended infiltration with glycerin, which will sublime during examination and bleed-off excess charge. Either of these methods reduce charging, but contaminate the microscope column. (*cf* Panessa and Gennaro, 1974, who note that the low vapor pressure of glycerin avoids excessive sublimation.)

During examination, charge accumulation is reduced by operating at minimal accelerating voltage (2.5–5 keV) and spot size, both of which serve to diminish the electron excitation volume and reduce radiation damage. The beam dwell-time during observation and recording must be minimal; always photograph images using the shortest time possible (≤30 sec). The maximum useful magnification in ideal situations is ~500 diameters; if higher magnification and/or resolution is desired, the specimens should be sputter coated.

The method of examining fresh-frozen, uncoated preparations was successfully used for insects by Nei (1974) and for plant surfaces by Ledbetter (1976). (Also see Falk *et al.*, 1970; Heslop–Harrison and Heslop–Harrison, 1970; and Mozingo *et al.*, 1970, for further observations on plant surfaces.) A major advantage of this method with botanical specimens is that the wax cuticle of the sample remains intact; the conventional reagents used for dehydration and critical-point drying dissolve the waxy coat (Murphy, 1978). Ledbetter's (1976) method for preparing and examining freshly frozen specimens is as follows: With a slow drying glue, attach the specimen to a cooled stub (McAlear and Germinario, 1973). Rapidly plunge the stub into liquid nitrogen (LN_2) for ~2 min, or until boiling stops, and immediately transfer into the microscope; retain visible, excess LN_2 over the specimen. Rapidly examine the specimen under the conditions noted above; after a while, collapse will become evident. Although Ledbetter (1976) successfully used accelerating voltages as high as 20 keV, the solid N_2 sublimes during vacuum exposure exactly as in the freeze-drying method, and thus examination must be rapid. If a cooled specimen stage is available, longer examination rates are possible (*e.g.*, Nei 1974; Koch, 1975; Echlin and Moreton, 1976; Echlin 1978). In general, magnifications of ~2000X are possible (Woods and Ledbetter, 1976).

The advantage of these methods is that they may rapidly be performed and evaluted. Nonfrozen materials may be sputter coated subsequently if the user is dissatisfied with preliminary results; typically, radiation damage effects will not be apparent at relatively low magnifications. Frozen materials usually cannot be salvaged, and one may need to resort to conventional preparation methods. Nevertheless, rapidly frozen specimens are probably truer to life than those which are chemically fixed. Despite the disadvantages noted above, success always affords a great deal of satisfaction. The aesthetic qualities of such micrographs has introduced SEM to many non-scientists as an art; many museums, including Chicago's Museum of Modern Art, have had exhibitions of SEM micrographs.

METALLIC IMPREGNATION

The use of heavy metal salts to increase the electron density of biological specimens, especially tissues, has been very successful. Most commonly the goal of metallic impregnation is to increase the general, overall density of the sample, although the identification of site-specific reactions deserves further research. In the former case the SEM image is composed of conventional secondary and backscattered electrons, whereas in the latter case one distinguishes the electron signals and confirms the presence of a metal-reactive site with x-ray microanalysis (*e.g.*, see Abraham and DeNee, 1974; Panessa and Gennaro, 1974; DeNee *et al.*, 1977; Becker and Sogard, 1979; Ogura and Laudate, 1980). This method becomes exciting when it is realized that these backscattered images closely resemble light microscope images, and excellent intermicroscope correlations may be made (Geissinger and Kamler, 1972; Kushida *et al.*, 1977; Okagi and Clark, 1977; Tannenbaum *et al.*, 1978; Ogura and Laudate, 1980). Another signal that has not yet been researched would be cathodoluminescence of site-selective stains which are both conductive and luminescent during electron exposure. Although cathodoluminescence studies have been conducted on biological specimens at this point no "stain" has been discovered which possesses luminescence and sufficient molecular weight for routine analyses. (See Pfefferkorn *et al.*, 1980, for a general review, and the bibliography by Brocker and Pfefferkorn, 1980.)

Another major application of metallic impregnation involves microdissection of the specimen during SEM examination, using micromanipulators; or different levels of the specimen are studied–the specimen is removed in order to expose another layer, and is reexamined. (Gold coatings are removed by cyanide treatment; see Sela and Boyde, 1977.) This would not be possible with coated specimens, unless they were recoated at each exposed level. A variety of tissues have been impregnated and dissected; the passage of blood cells through bone marrow was studied by Irino *et al.* (1975); spleen was studied by Fujita (1974), Irino *et al.* (1977 and 1978) and Murakami *et al.* (1977); kidney by Tokunaga *et al.* (1974), Murakami and Jones (1980); liver by Itoshima *et al.* (1974); and plant materials by Kunoh *et al.* (1975).

Metallic impregnation methods may be classified as one-step or ligand-mediated binding: In the first situation, the tissue is exposed only once to a heavy metal salt; while in the latter tissue reactivity toward metal salts is increased by exposure to a ligand, and the tissue subsequently exposed to the metal solution. Usually the latter is repeated several times, resulting in a cascade effect of tissue-ligand-metal-ligand, etc. As expected, the latter more effectively enhances electron density than single metal-salt application.

Early SEM researchers applied various heavy metal salts to tissues for an increase in conductivity; for example, potassium iodide (Jacques *et al.*, 1965), potassium iodide and lead acetate (Panessa and Gennaro, 1973; Rodman and Caughey, 1974), potassium permanganate (Nowell *et al.*, 1972), uranyl nitrate (Idle, 1971), and uranyl acetate (Mumaw *et al.*, 1976). However, conductivity was not enhanced sufficiently for resolutions better than those obtained with untreated specimens.

A different approach was then attempted by taking advantage of the osmium tetroxide postfixation normally used in preparation. Seligman *et al.* (1966) found that tissue reactivity toward osmium was increased by exposure to thiocarbohydrazide (TCH): Following postfixation in osmium, tissues were washed with water, exposed to TCH, again washed, and then re-exposed to osmium. Thus, this method is referred to as the OTO method. A significant increase in the contrast of TEM specimens resulted. Kelley *et al.* (1973) applied this technique to biological specimens for SEM study, but achieved only low resolution and magnification. Subsequently, an additional exposure to TCH and osmium (*i.e.*, OTOTO) was introduced by Malick and Wilson (1975a,b) for animal tissues and by Postek and Tucker (1977) for plant specimens. Magnification was increased to ~30,000X, and significantly higher resolution was attained because increased accelerating voltages could be used. Cryofracturing of OTOTO-prepared specimens for examining intracellular structure, has been very successful with plant tissues (Woods and Ledbetter, 1976) and animal tissues (Munger and Mumaw, 1976).

Concurrently with the work cited above, Murakami (1973, 1974, and 1978) exposed tissue specimens to tannic acid, either during or after primary glutaraldehyde fixation. Postosmication was enhanced by the tannin treatment: Simionescu and Simionescu (1976a,b) have completely discussed the mordant effect of tannin (syn: galloylglucose). The original tannin–osmium method (abbreviated TAO) has been modified by Takahashi (1979) into the O-GTA-O-GTA-O (osmium-glutaraldehyde/tannin, *etc.*), which permits repeated imaging of the same specimen at 20 keV and very high magnifications of ~200,000X (Murphy, 1980). These methods require considerable preparation time, however, and thus Sweney and Shapiro (1977) and Sweney *et al.* (1979) introduced the GTA-O-TA-O method for rapid preparations. The most recent modification of the original TAO method is by Murakami and Jones (1980), who use both tannin and thiocarbohydrazide as mordants in the TAOTO method (tannin, osmium, TCH, osmium). This technique provides the best imaging qualities—nearly comparable to coated specimens—and also reduces the shrinkage artifacts inherent during organic dehydration and critical-point drying.

Another osmium-based metallic impregnation method uses osmium tetramethylethylenediamine (OsTMEDA) as a substitute for osmium tetroxide postfixation (Wilson *et al.*, 1979). By far, this is the most rapid technique available; the substitution of OsTMEDA for OsO_4 during normal processing (glutaraldehyde fixation, OsTMEDA, ethanol dehydration, and critical-point drying) is the only required treatment. The results of this method are comparable to the original TAO and OTOTO techniques, but does not reach the resolution obtainable with the TAOTO method (Murphy, 1980).

Moderately site-specific staining using silver was introduced by Geissinger (1972; Geissinger and Kamler, 1972) for increasing electron conductivity and directly correlating SEM and light-microscope images. Goldman and Leif (1973) used silver nitrate to increase general specimen conductivity, and obtained good results at 20 keV and 40,000X magnification. Silver has been used for backscattered imaging (Vogel *et al.*, 1976; Tannenbaum *et al.*, 1978), and will probably become very important as pathologists desire intermicroscope correlations (*e.g.*, see Carter, 1980). Because silver is not as commonly used as tannin or TCH methods, it will not be discussed further. Refer to Geissinger (1974) for detailed methods.

Other site-selective reactions, which do not increase over-all density, are used for identifying animal immunoprotein complexes or plant lecithin reactions. Certain heavy metals may be used to "trace" or "mark" selected chemical species; the metal may then be detected by x-ray microanalysis, and the reaction site noted. The most commonly used marker is gold (Horisberger, 1979; Horisberger and Rosset, 1977; Horisberger *et al.*, 1975), with platinum (Goodman *et al.*, 1979) and platinum/palladium (Hoyer *et al.*, 1979) also being investigated. More complete reviews of selective markers are available in Molday (1977), Hodges and Hallowes (1979), and Goodman *et al.* (1980). It is stressed that these methods do not render the specimen conductive: an evaporated or sputtered coating is required (Echlin and Kaye, 1979; Peters, 1979).

Osmium-Thiocarbohydrazide Methods

The reactions of osmium tetroxide with cellular macromolecules was discussed in Chapter 4 and will not be repeated. The uptake of osmium by the specimen is enhanced by treatment with TCH (Seligman *et al.*, 1965, 1966, and 1968). The OTO method follows this sequence: osmium postfixation ⟶ water wash ⟶ 1% aqueous TCH ⟶ 1% aqueous osimum ⟶ water wash ⟶ dehydration (Kelley *et al.*, 1973 and 1975). TCH must be carefully prepared to avoid precipitation artifacts; the following method devised by Woods and Ledbetter (1976) is recommended:

1. Place 0.5 gm of crystalline TCH in a beaker and repeatedly rinse with distilled water until the originally pink crystals are colorless.
2. Decant the excess water and resuspend the cleaned TCH in 25 ml of distilled water, heat to 60°C until dissolution occurs (a few minutes), then gradually cool to room temperature (~1 hour).
3. Filter the solution through 0.47 μm Millipore filter.
4. Store away from light and refilter before use.

Numerous researchers have used the OTO method for various tissues, but resolution is not significantly better than that obtained with uncoated or only osmicated tissues (*e.g.*, Nowell *et al.*, 1972; Kelley *et al.*, 1973 and 1975; Waterman, 1974; Woods and Ledbetter, 1974; Munger and Mumaw, 1976; Jones, 1978; Siew, 1978).

However, if the tissue is processed *via* OTOTO–*i.e.*, a triple exposure to osmium–conductivity and resolution are enhanced and higher accelerating voltages may be used (Malick and Wilson, 1975a,b). The preparation sequence follows osmium postfixation ⟶ water wash ⟶ 1% aqueous TCH ⟶ water wash ⟶ 1% aqueous osmium ⟶ water wash ⟶ 1% aqueous TCH ⟶ water wash ⟶ 1% aqueous osmium ⟶ water wash ⟶ dehydration. Each water wash must be very thorough to avoid surface contamination by precipitation of OsO_4 and TCH;

Malick and Wilson (1975a) recommend six changes of water over a period of 15 min at each wash step. The TCH exposures are for 20–30 min each, and osmication is for 2–3 hrs with agitation. The entire process is carried out at room temperature. Tissues processed by the OTOTO method may be examined at 20 keV and 20,000X with good resolution. The higher accelerating voltage permits a greater depth of penetration and thus allows observation of underlying cells, whereas surfaces are observed with lower (5–10 keV) accelerating voltages. The OTOTO method has been successfully used with animal tissues (Kelley *et al.* 1975; Malick and Wilson, 1975a,b) and botanical specimens (Postek and Tucker, 1977). Estable-Puig *et al.* (1976) note that these specimens are most amenable to intermicroscope (SEM and TEM) correlations: Following SEM examination, the specimen is infiltrated with an epoxy resin, polymerized, sectioned, and examined in the TEM.

Specimens prepared as above are ultimately critical-point dried for the examination of cell surfaces. The direct observation of intracellular structures, such as organelles, requires cryofracturing of OTO- or OTOTO-treated specimens; *i.e.*, the specimens are fractured, then critical-point dried. Munger and Mumaw (1976) followed the OTO procedure (Murphy, 1978, notes that OTOTO would be better), and dehydrated with a graded series of ethanol followed by Freon 113. The bulk specimen is positioned between two LN_2-precooled copper plates, and then fractured by tapping the plate with a metal rod. Finally, the specimens are infiltrated with cool Freon 113 and critical-point dried. Examination of the fracture surfaces is as in the OTO method. Munger and Mumaw (1976) also enhanced the data signal from nucleic acid-containing organelles (*e.g.*, nuclei) by *en bloc* staining/teritiary fixation with uranyl acetate.

A much longer procedure for cryofracturing specimens was introduced by Woods and Ledbetter (1976). Referred to as the OTOTO-cryoresinfracture method, the specimens are infiltrated and embedded in an epoxy resin following OTOTO treatment. The embedded specimen is frozen in LN_2 and fractured; the resin is then dissolved; and the specimen examined. This method requires approximately four days of preparation time, and therefore is not recommended. The detailed methodology may be found in Woods and Ledbetter (1976) and Murphy (1978). Although this method provides higher resolution than the OTO-cryofracture technique, this author believes it is due to the enhanced contrast of OTOTO, which is the only advantage. Simply modify the Munger and Mumaw (1976) technique to OTOTO-cryofracture.

Regardless of which method is used, success depends, first, upon working with fresh, filtered TCH solutions and very thoroughly washing the specimens between osmium and TCH exposures. Stock solutions of TCH form a flake-like, white precipitate: always prepare fresh solution and filter immediately before use. Excess, unreacted osmium or TCH must be removed from the tissue to avoid precipitation at the specimen surface; rinse many times (approximately six) with constant agitation. As an aside, early researchers (*e.g.*, Kelley *et al.*, 1975) recommended an ambient reaction temperature of 60°C, but it has subsequently been shown that room temperature does not adversely affect the reaction (*e.g.*, Malick and Wilson, 1975a). To ensure that the reaction occurs throughout the bulk specimen, rather than only at its surface, frequently agitate the specimens. Finally, all precautions concerning handling osmium tetroxide must be enforced.

Tannin-Osmium Methods

Concurrent with the development of TCH methods, Murakami (1973) introduced tannin to enhance tissue reactivity toward osmium. Subsequent experimentation and modification of the original TAO (tannin–osmium) method by Murakami (1974; Murakami *et al.*, 1977) was very successfully used for the microdissection of tissues within the SEM (Fujita, 1974; Itoshima, *et al.*, 1974 and 1978; Tokunaga *et al.*, 1974; Irino *et al.*, 1975, 1977, and 1978; Kunoh *et al.*, 1975; and Murakami *et al.*, 1977). The revised method, conducted at room temperature, is as follows (Murakami *et al.*, 1977):

1. Primary fixation is with 2.5% buffered glutaraldehyde containing 0.1 *M* PO_4.
2. Treat the specimens with an aqueous mixture of 2% guanadine hydrochloride and 2% tannic acid for 8 hr.
3. Wash in several changes of distilled water for 30 min.
4. Immerse in 2% osmium tetroxide (aqueous) for 8 hr; agitate frequently.

5. Dehydrate in ethanol, infiltrate with amyl acetate, and critical-point dry.

Specimens prepared in this manner show high contrast and resolution at 25 keV, and magnifications up to 30,000×.

A significantly more rapid procedure employing tannin was introduced by Sweney and Shapiro (1977; also see Sweney *et al.*, 1979) for specimens as diverse as insects, animal tissues, and botanical specimens. The preparation sequence in this method is as follows: glutaraldehyde fixation ⟶ glutaraldehyde/tannic acid ⟶ buffer wash ⟶ buffered OsO_4 ⟶ water wash ⟶ tannic acid ⟶ water wash ⟶ OsO_4 ⟶ water wash. This is conveniently abbreviated as GTA-O-TA-O. The actual preparation of specimens is as follows (Sweney and Shapiro, 1977; Murphy, 1980):

1. Primary fixation is with 1% glutaraldehyde buffered with 1.0 *M* cacodylate or phosphate for 30 min. This and subsequent steps are conducted at room temperature.
2. Transfer to a filtered mixture of 3% glutaraldehyde and 8% tannic acid in buffer for 0.5–12 hr (large specimens require long durations).
3. Rinse in buffer for 5 min, several changes.
4. Postfix in 0.5% OsO_4 in cacodylate (or in phosphate) buffer for 30 min with agitation.
5. Rinse six times with distilled water for a total of 15 min.
6. Submerge in fresh, aqueous 0.5% tannic acid for 30–60 min. Agitate.
7. Wash three times with distilled water for a total of 5 min.
8. Re-expose to aqueous 0.5% OsO_4 for 30–60 min. Agitate.
9. Rinse six times with distilled water for a total of 15 min.
10. Dehydrate in ethanol, infiltrate with ethanol/Freon TF, and critical-point dry with Freon 13.

Specimens are examined as in the TAO method. This technique is very useful because it is rapid; but repeatable success requires attention to a few parameters (Sweney and Shapiro, 1977): the fixative/tannic acid must be filtered just prior to use to avoid precipitation. Agitation is essential for rapid and thorough penetration of the tissue during exposure to tannic acid and osmium tetroxide. Washing must also be complete to avoid precipitation artifacts at the specimen surface. Finally, these researchers note that some tissues are sufficiently strengthened to tolerate air-drying, although critical-point drying is standard. Murphy (1980) includes a procedure involving vascular perfusion during primary fixation.

Another modification of the tannin-osmium method was introduced by Takahashi (1977, 1979). It is the O-GTA-O-GTA-O method (*i.e.*, osmium ⟶ glutaraldehyde/tannic acid ⟶ osmium, etc.). The highest magnification employed was 120,000X at 20 keV, with excellent resolution. When tissues are to be fixed by immersion, the following sequence applies (Takahashi, 1979; Murphy, 1980):

1. Fix tissues in 2% OsO_4 in 0.1 *M* buffer at 4°C.
2. Postfix tissues with a mixture of buffered 8% glutaraldehyde and 2% tannic acid for 12 hr at 4°C. Change the solution three times.
3. Rinse three times in buffer for a total of 15 min, cold.
4. Immerse in 2% buffered OsO_4 for 2 hr at 4°C.
5. Rinse three times in buffer for a total of 15 min.
6. Repeat steps 2–5.
7. Dehydrate and critical-point dry.

The procedure for vascularly perfused specimens is as follows (Takahashi, 1979; Murphy, 1980):

1. Preperfuse with Ringer's solution, warm.
2. Perfuse with 1.5% buffered glutaraldehyde for 5 min, warm.
3. Continue perfusion with a chilled mixture of buffered 1.5% glutaraldehyde and 0.5% tannic acid for 15 min.
4. Complete perfusion with a chilled mixture of buffered 1.5% glutaraldehyde and 1% tannic acid for 10 min.
5. Remove tissue, mince, and wash with several changes of cold buffer for 1 hr.
6. Continue with immersion fixation as given above.

The same precautions as noted for the method of Sweney and Shapiro (1977) apply with this technique.

Murakami very recently modified his earliest work and has combined tannin and thiocarbohydrazide treatments into the TAOTO method (*i.e.*, glutaraldehyde/tannin ⟶ osmium ⟶ TCH ⟶ osmium, with intermediate rinses) (Murakami and Jones, 1980). Good resolution and contrast at 90,000X and 25 keV were noted; in addition, shrinkage during dehydration and critical-point drying was significantly reduced relative to normal tissue dimensions. They prepared rat kidney as follows:

1. The kidney was preperfused with Ringer's solution, followed by perfusion fixation with phosphate buffered (0.1 *M*) glutaraldehyde (2%).
2. Tissue slices were removed and immersion fixed for 4 hr.
3. The specimens were exposed to a buffered mixture of 2% tannic acid and 2% glutaraldehyde for 4 hr.
4. Washing with buffer was for 8 hr.
5. The tissues were exposed to buffered 2% OsO_4 for 3 hr.
6. Washing with distilled water was for 6 hr.
7. Tissues were exposed to aqueous 1% TCH for 2 hr.
8. Washing with distilled water was for 4 hr.
9. Tissues were stained with aqueous 1% OsO_4 for 3 hr.
10. Washing with distilled water was for 6 hr.
11. Ethanol dehydration and CO_2 critical-point drying complete the method.

Although this procedure is very lengthy, Murakami and Jones (1980) published very fine micrographs. It is likely that future experimentation will significantly lower the amount of time required for processing. Again, the precautions noted earlier must be noted for success.

Osmium Tetramethylethylenediamine (OsTMEDA) Method

Wilson *et al.* (1979) introduced a new method very different from those discussed above. Whereas all preparatory steps for primary fixation; dehydration, and critical-point drying remain standard, postfixation is with osmium tetramethylethylenediamine (OsTMEDA), rather than osmium tetroxide. The postfixation vehicle contains acetic acid, not the conventional buffers. The OsTMEDA method is as follows (Wilson *et al.*, 1979):

1. Primary fixation with buffered 3% glutaraldehyde is for 3 hrs.
2. Wash three times with buffer for a total of 15 min.
3. Post-fix with 0.5% OsTMEDA in 3% acetic acid for 3 hrs. The pH must be held at 7.2 to avoid contamination.
4. Wash three times with buffer for a total of 15 min. Do not exceed this time.
5. Dehydrate in ethanol and critical point dry in Freon 13.

Good resolution was noted with 30 keV at magnification 40,000X. This technique has the advantage that fairly rapid processing is possible. On the other hand, commercially purchased OsTMEDA is expensive (~$30/100 mg).

SUMMARY

The advantages and disadvantages of metallic impregnation have been discussed by Murphy (1978 and 1980). It must be understood that the experimental requirements determine the optimal method for enhancing electron density. For example, tissues which are to be dissected should be uncoated for examination, while metal coating is most amenable to strict surface-morphology analyses. When correlative microscopy (TEM and SEM) of the same specimen is desired, uncoated materials are again preferred. As discussed in previous chapters, as the reliability (*i.e.*, reduction of thermal damage and contamination) of evaporated and sputtered films increases, methods such as metallic impregnation need not be universally applied. In short, determine the objectives of the study and then select the best method.

REFERENCES

Abraham, J. L. and P. B. DeNee (1974). Biomedical application of backscattered electron imaging–one year's experience with SEM histochemistry. *IITRI/SEM* p. 251.

Becker, R. P. and M. Sogard (1979). Visualization of subsurface structures in cells and tissues by backscattered electron imaging. *SEM, Inc.* **2**:835.

Brocker, W. and G. Pfefferkorn (1980). Bibliography on cathodoluminescence, part II. *SEM, Inc.* **1**:298.

Carter, H. W. (1980). Clinical applications of scanning electron microscopy (SEM) in North America with SEM's role in comparative microscopy. *SEM, Inc.* **3**:115.

—— *et al.* (1977). Heavy metal staining of paraffin, epoxy, and glycol methacrylate embedded biological tissue for scanning electronmicroscope histology. *IITRI/SEM* **2**:83.

DeNee, P. B. and E. R. Walker (1975). Specimen coating technique for the SEM: a comparative study. *IITRI/SEM* p. 225.

Echlin, P. (1978). Low temperature scanning electron microscopy: a review. *J. Micros.* **112**:47.

—— and G. Kaye (1979). Thin films for high resolution conventional scanning electron microscopy. *SEM, Inc.* **2**:21.

—— and R. Moreton (1976). Low temperature techniques for SEM. *IITRI/SEM* **1**:753.

Estable–Puig, R. F. de *et al.* (1976). SEM study of brain macrophages with quantitative data on cell surface areas and volume. *IITRI/SEM* **2**:196.

Falk, R., E. Gifford and E. Cutter (1970). SEM of developing plant organs. *Science* **168**:1471.

Fujita, T. (1974). A SEM study of the human spleen. *Arch. Histol. Jap.* **37**:187.

Geissinger, H. D (1972). The use of silver nitrate as a stain for scanning electron microscopy of arterial intima and paraffin sections of kidney. *J. Micros.* **95**:471.

—— (1974). Silver as a stain. In: *Principles and Techniques of Scanning Electron Microscopy*, Vol. 2, M. A. Hayat, ed. Van Nostrand Reinhold, New York, p. 26.

—— and H. Kamler (1972). Precise and fast correlation of light microscopic and scanning electron microscope images. *Car. Res. Dev.* **5**:13.

Goldman, M. A. and R. C. Leif (1973). A wet chemical method for rendering SEM samples conductive and observations on the surface morphology of human erythrocytes and Ehrlich ascites cells. *Proc. Nat. Acad. Sci.* **70**:3599.

Goodman, S. L. *et al.* (1980). A review of the collodial gold marker system. *SEM, Inc.* **2**:133.

—— *et al.* (1979). Collodial gold probes–a further evaluation. *SEM, Inc.* **3**:619.

Heslop-Harrison, Y. and J. Heslop-Harrison (1970). SEM of fresh leaves of *Pinguicula. Science* **167**:172.

Hodges. G. M. and R. C. Hallowes, editors (1979). *Biomedical Research Applications of Scanning Electron Microscopy.* Academic Press, New York.

Horisberger, M. (1979). Evaluation of colloidal gold as a cytochemical marker for transmission and scanning electron microscopy. *Biol. Cellulaire* **36**:253.

—— and J. Rosset (1977). Gold granules, a useful marker for SEM. *IITRI/SEM* **2**:75.

—— *et al.* (1975). Colloidal gold granules as markers for cell surface receptors in the scanning electron microscope. *Experientia* **31**:1147.

Howden, H. F. and L. E. C. Ling (1974). Low magnification study of uncoated specimens. In: *Principles and Techniques of Scanning Electron Microscopy*, Vol. 1, M. A. Hayat, ed. Van Nostrand Reinhold, New York, p. 149.

Hoyer, L. C. *et al.* (1979). Scanning immunoelectron microscopy for the identification and mapping of two or more antigens on cell surfaces. *SEM, Inc.* **3**:629.

Idle, D. (1971). Preparation of plant material for SEM. *J. Micros.* **93**:77.
Irino, S. *et al.* (1977). Open circulation in the human spleen. Dissection SEM of conductive-stained tissue and observation of resin vascular casts. *Arch. Histol. Jap.* **40**:297.
—— *et al.* (1978). Microdissection of tannin–osmium impregnated specimens in the scanning electron microscope: demonstration of arterial terminals in human spleen. *SEM, Inc.* **2**:111.
—— *et al.* (1975). SEM studies of microvascular architecture, sinus wall, and transmural passage of blood cells in the bone marrow by a new method of injection replica and non-coated specimens. *IITRI/SEM* p. 267.
Itoshima, T. *et al.* (1974). Fenestrated endothelium of the liver sinusoids of the guinea pig as revealed by SEM. *Arch. Histol. Jap.* **37**:15.
—— *et al.* (1978). Alterations of liver cell nuclei in hepatitis as revealed by scanning electron microscopy. *SEM, Inc.* **2**:203.
Jacques, W. E. *et al.* (1965). Application of the SEM to human tissues: a preliminary study. *Exp. Mol. Pathol.* **4**:576.
Jones, J. B. (1978). Scanning electron microscopy of human hypertensive renal disease. *SEM, Inc.* **2**:937.
Kelley, R. O. *et al.* (1973). Ligand-mediated osmium binding: its application in coating biological specimens for scanning electron microscopy. *J. Ultrastr. Res.* **45**:254.
—— *et al.* (1975). TCH-mediated Os binding: a technique for protecting soft biological specimens in the SEM. In: *Principles and Techniques of Scanning Electron Microscopy*, Vol. 4, M. A. Hayat, ed. Van Nostrand Reinhold, New York, p. 34.
Koch, G. R. (1975). Preparation and examination of specimens at low temperature. In: *Principles and Techniques of Scanning Electron Microscopy*, Vol. 4, M. A. Hayat, ed. Van Nostrand Reinhold, New York, p. 1.
Kunoh, H. *et al.* (1975). The conductive staining method suitable for x-ray microanalysis and micromanipulation of plant materials. *J. Elec. Micros.* **24**:301.
Kushida, T. *et al.* (1977). Observation on the same place in semi-thin section with both light and electron microscopy. *J. Elec. Micros.* **26**:345.
Ledbetter, M. C. (1976). Practical problems in observation of unfixed, uncoated plant surfaces by SEM. *IITRI/SEM* **2**:453.
Malick, L. E. and R. B. Wilson (1975a). Evaluation of a modified technique for the SEM examination of vertebrate specimens without evaporated metal layers. *IITRI/SEM* p. 259.
—— and R. B. Wilson (1975b). Modified TCH procedure for SEM: routine use for normal, pathological, or experimental tissues. *Stain Technol.* **50**:265.
McAlear, J. H. and L. T. Germinario (1973). The cold stub: a simple stage for the direct observation of frozen samples in the SEM. *Proc. 31st Am. EMSA Meet.*, p. 450.
Molday, R. S. (1977). Cell surface labeling techniques for SEM. *IITRI/SEM* **2**:59.
Mozingo, H. *et al.* (1970). "Venus" flytrap observations by SEM. *Am. J. Bot.* **57**:593.

Mumaw, V. R. *et al.* (1976). The use of lead and uranyl ions as a stain in scanning electron microscopy. *Proc. 34th Ann. EMSA Meet.*, p. 314.

Munger, B. L. (1977). The problem of specimen conductivity in electron microscopy. *IITRI/SEM* **1**:481.

—— and V. Mumaw (1976). Specimen preparation for SEM study of cells and cell organelles in uncoated preparations. *IITRI/SEM* **1**:275.

Murakami, T. (1973). A metal impregnation method of biological specimens for scanning electron microscopy. *Arch. Histol. Jap.* **35**:323.

—— (1974). A revised tannin–osmium method for non-coated scanning electron microscope specimens. *Arch. Histol. Jap.* **36**:189.

—— (1978). Tannin–osmium conductive staining of biological specimens for non-coated scanning electron microscopy. *Scanning* **1**:127.

—— and A. L. Jones (1980). Conductive staining of biological specimens for non-coated scanning electron microscopy: double staining by tannin–osmium and osmium–thiocarbohydrazide–osmium methods. *SEM, Inc.* **1**:221.

—— *et al.* (1977). Modified TAO conductive staining method for noncoated SEM specimens. Its application to microdissection SEM of the spleen. *Arch. Histol. Jap.* **40**:35.

Murphy, J. A. (1978). Non-coating techniques to render biological specimens conductive. *SEM, Inc.* **2**:175.

—— (1980). Non-coating techniques to render biological specimens conductive/1980 update. *SEM, Inc.* **1**:209.

Nei, T. (1974). Cryotechniques. In: *Principles and Techniques of Scanning Electron Microscopy*, Vol. 1, M. A. Hayat, ed. Van Nostrand Reinhold, New York, p. 113.

Nowell, J. A. *et al.* (1972). Stabilization and replication of soft tubular and alveolar systems—a scanning electron microscope study of the lung. *IITRI/SEM* p. 305.

Ogura, K. and A. Laudate (1980). Comparative observation with a light microscope and an SEM in backscattered electron mode. *SEM, Inc.* **1**:233.

Okagi, T. and B. Clark (1977). X-ray microprobe analysis of biological stains using ultra thin paraffin embedded sections. *IITRI/SEM* **2**:153.

Panessa, B. J. and J. F. Gennaro, Jr. (1973). Use of potassium iodide/lead acetate for examining uncoated specimens. *IITRI/SEM* p. 395.

—— and J. F. Gennaro, Jr. (1974). Intracellular structures. In: *Principles and Techniques of Scanning Electron Microscopy*, Vol. 1, M. A. Hayat, ed. Van Nostrand Reinhold, New York, p. 226.

Peters, K. R. (1979). Scanning electron microscopy at macromolecular resolution in low energy mode on biological specimens coated with ultrathin metal films. *SEM, Inc.* 2:133.

Pfefferkorn, G., *et al.* (1980). The cathodluminescence method in the scanning electron microscope. *SEM, Inc.* **1**:251.

Postek, M. T. and S. C. Tucker (1977). TCH binding for botanical specimens for SEM: a modification. *J. Micros.* **110**:71.

Rodman, N. F. and R. C. Caughey (1974). Effects of specimen preparation techniques on surface ultrastructural study of platelet aggregates. *IITRI/SEM* p. 306.

Sela, J. and A. Boyde (1977). Cyanide removal of Au from SEM specimens. *J. Micros.* **111**:229.

Seligman, A. M. *et al.* (1965). Histochemical demonstration of some oxidized macromolecules with thiocarbohydrazide (TCH) or thiosemicarbazide (TSC) and osmium tetroxide. *J. Histochem. Cytochem.* **13**:629.

—— *et al.* (1968). Osmium containing compound with multiple basic or acidic groups as stains for ultrastructure. *J. Histochem. Cytochem.* **16**:87.

—— *et al.* (1966). A new staining method (OTO) for enhancing contrast of lipid-containing membranes and droplets in osmium tetroxide fixed tissue with osmiophilia thiocarbohydrazide (TCH). *J. Cell Biol.* **30**:424.

Siew, S. (1978). Scanning electron microscopy of acute rheumatic valvulitis. *SEM, Inc.* 2:341.

Simionescu, N. and M. Simionescu (1976a). Galloylglucoses of low molecular weight as mordants in electron microscopy. I. Procedure and evidence for mordanting effect. *J. Cell Biol.* **70**:608.

—— and M. Simionescu (1976b). Galloylglucoses of low molecular weight as mordants in electron microscopy. II. The moiety and functional groups possibly involved in the mordanting effect. *J. Cell Biol.* **70**:622.

Sweney, L. R. *et al.* (1979). SEM of uncoated human metaphase chromosomes. *J. Micros.* **115**:151.

—— and B. L. Shapiro (1977). Rapid preparation of uncoated biological specimens for scanning electron microscopy. *Stain Technol.* **52**:221.

Takahashi, G. (1977). (Glutaraldehyde)–OsO_4–tannic acid–glutaraldehyde–OsO_4 fixation and staining of biological specimens for electron microscopy. *J. Electron Micros.* **26**:326.

—— (1979). Conductive staining method. *The Cell* **11(4)**:114.

Tannenbaum, M. *et al.* (1978). SEM, BEI, and TEM ultrastructural characteristics of normal preneoplastic, and neoplastic human transitional epithelium. *SEM, Inc.* 2:949.

Tokunaga, J. *et al.* (1974). Freeze cracking of SEM specimens. A study of the kidney and spleen. *Arch. Histol. Jap.* **37**:165.

Vogel, G. *et al.* (1976). Backscattered electron imaging of nuclei in rat testis. *IITRI/SEM* 2:410.

Waterman, R. E. (1974). SEM study of early facial development in rodent and man. *IITRI/SEM* p. 533.

Wilson, D. C. *et al.* (1979). A simple conductive coating for SEM of cells and tissues. *Proc. 37th Ann. EMSA Meet.*, p. 360.

Woods, P. S. and M. C. Ledbetter (1974). A method of direct visualization of plant cell organelles for scanning electron microscopy. *Proc. 32nd Ann. EMSA Meet.*, p. 122.

—— and M. C. Ledbetter (1976). Cell organelles at uncoated cryofractured surfaces as viewed with the SEM. *J. Cell Sci.* **21**:47.

Appendix

CACODYLATE BUFFERS AND FIXATIVES

A. Stock Solution: 0.4 *M* Cacodylate Buffer

$Na(CH_3)_2AsO_2 \cdot 3H_2O$	42.8	gm
Distilled water to make	500	ml

If sodium cacodylate is in anhydrous form, use 32.0 gm and make up to 500 ml.

B. Working Solution: 0.2 *M* Cacodylate Buffer

Stock 0.4 *M* Buffer	50	ml	
0.2 *M* HCl	5.4	ml → pH 7.4 }	Select the
	8.4	ml → pH 7.2 }	appropriate
	12.6	ml → pH 7.0 }	pH for speci-
	18.6	ml → pH 6.8 }	men type.
Distilled water to make	100	ml	

C. Cacodylate Buffered Glutaraldehyde (2.5%)

Working 0.2 *M* Buffer	50	ml
25% Glutaraldehyde (aqueous)	10	ml
Distilled water to make	100	ml

The final concentration of fixative is 2.5% in 0.1 *M* buffer. If desired, adjust osmolarity by adding sucrose, glucose, or sodium chloride.

D. Cacodylate Buffered Formaldehyde (3%)

Working 0.2 *M* Buffer	50	ml
40% Paraformaldehyde (aqueous)	7.5	ml
Distilled water to make	100	ml

E. Cacodylate Buffered (0.1 *M*) Formaldehyde (2%)/Glutaraldehyde (2.5%)

Refer to "Paraformaldehyde: Preparation of Aqueous Stock Solution."

Working 0.2 *M* Buffer	50	ml
10% Paraformaldehyde (aqueous)	20	ml
25% Glutaraldehyde (aqueous)	10	ml
Distilled water to make	100	ml

F. Cacodylate (0.1 *M*) Buffered Acrolein (10%)

Working 0.2 *M* Buffer	50	ml
Acrolein	10	ml
Distilled water to make	100	ml

G. Cacodylate Buffered (0.1 *M*) Acrolein (1%)/Glutaraldehyde (2.5%)

Working 0.2 *M* Buffer	50	ml
Acrolein	1	ml
25% Glutaraldehyde (aqueous)	10	ml
Distilled water to make	100	ml

H. Cacodylate Buffered (0.1 *M*) Osmium Tetroxide (2%)

Working 0.2 *M* Buffer	4	ml
4% OsO_4 (aqueous)	2	ml

I. Cacodylate Buffered (0.1 *M*) Glutaraldehyde (0.8%)/Osmium Tetroxide (0.7%)

Prepare 0.2 *M* cacodylate buffer at pH 7.4.

Prepare 2.5% glutaraldehyde and 10% osmium tetroxide in 0.1 *M* cacodylate buffer.

Cool to 0–4°C in an ice bath and combine the above fixatives, 1:2 of glutaraldehyde : osmium tetroxide.

Notes

1. Arsenic-containing solutions are toxic and must be carefully handled.
2. The pH of the working buffer will drop over time; readjust to desired pH with 0.2 *M* HCl.
3. Refrigerate all solutions in glass-stoppered vessels.

Reference

Sabatini, D. D. *et al.* (1963). New means of fixation for electron microscopy and histochemistry. *Anat. Rec.* **142**:274.

COLLIDINE BUFFER AND FIXATIVES

A. Stock Solution

S-Collidine (pure)	5.34	gm
Distilled water to make	100	ml

B. Working Solution

Stock Buffer Solution	50	ml
1.0 *N* HCl	9	ml (adjust volume for desired pH)
Distilled water to make	100	ml

C. Collidine-Buffered OsO_4 (2%)

Working Buffer Solution	20	ml
4% OsO_4 (aqueous)	10	ml

D. Collidine-Buffered (0.2 *M*) Formaldehyde (10%)

1. Preparation of Formaldehyde

Paraformaldehyde powder	10	gm
Distilled water	70	ml

Heat the above to 70°C for 20 min. with stirring. Add ∿6 drops 1 *N* NaOH, while stirring, until the solution clears.

2. Final Solution

Formaldehyde	70	ml
S-Collidine (pure)	2.4	ml
1 *N* HCl	5.0	ml
Distilled water to make	100	ml

Notes

1. The stock and working solutions are stable indefinitely at room temperature.
2. Collidine buffered glutaraldehyde may be prepared.

Reference

Bennett, H. S. and J. H. Luft (1959). S-Collidine as a basis for buffering fixatives. *J. Biophys. Biochem. Cytol.* **6**:113.

KELLENBERGER'S FIXATIVE FOR BACTERIA: OSMIUM TETROXIDE AND URANYL ACETATE

A. Stock Veronal Acetate Buffer

Sodium veronal	2.94 gm
Sodium acetate (hydrated)	1.94 gm
Sodium chloride	3.40 gm
Distilled water to make	100 ml

B. Working Kellenberger Buffer

Stock veronal acetate buffer	5.0 ml
Distilled water	13.0 ml
0.1 *N* HCl	7.0 ml
1.0 *M* $CaCl_2$	0.25 ml

The desired pH is 6.0. Prepare just prior to use to avoid microorganism contamination.

C. Kellenberger's OsO_4 Fixative (1%)

Kellenberger Buffer	8 ml
4% OsO_4 (aqueous)	2 ml

D. Washing Fixative: Uranyl Acetate (0.5%)

Kellenberger Buffer	10 ml
Uranyl acetate	0.05 gm

E. Tryptone Medium

Bacto-Tryptone (Difco)	1.0 gm
NaCl	0.5 gm
Distilled water	100 ml

Reference

Ryter, A. and E. Kellenberger (1958). Etude au microscope électronique de plasma contenant de l'acid désoxyribonucleique. I. Les nucléotides des bactéries en croissance active. *Z. Naturf.* **13**:597.

SORENSON'S SODIUM PHOSPHATE BUFFER: FORMULATION I (0.1 *M*)

A. Stock Solution: 0.2 *M* Phosphate Buffer

1. Preparation of 0.2 *M* sodium phosphate monobasic

$Na_2HPO_4 \cdot 7H_2O$	26.85 gm
Distilled water to make	500 ml

2. Preparation of 0.2 *M* sodium phosphate dibasic

$NaH_2PO_4 \cdot H_2O$	13.80 gm
Distilled water to make	500 ml

B. Working Solution: 0.1 *M* Phosphate Buffer

The following volumes of sodium phosphate monobasic and dibasic are combined for the desired pH. The mixture is then made up to 100 ml with distilled water.

	0.2 *M* SODIUM PHOSPHATE	
pH	MONOBASIC	DIBASIC
6.8	25.5 ml	24.5 ml
7.0	19.5 ml	30.5 ml
7.2	14.0 ml	36.0 ml
7.4	9.5 ml	40.5 ml

The osmolarity at pH 7.2 is 226 m*OsM*. Adding 0.18 *M* sucrose to 0.1 *M* working buffer increases osmolarity to 425 m*OsM*.

C. Phosphate Buffered Glutaraldehyde (2%)

Working 0.1 *M* Solution	100 ml
25% Glutaraldehyde (aqueous)	8 ml

D. Phosphate Buffered Osmium Tetroxide (2%)

Working 0.1 *M* Solution	4 ml
4% OsO_4 (aqueous)	2 ml

E. Phosphate Buffered Formaldehyde (3%)

Working 0.1 *M* Solution	50 ml
40% Paraformaldehyde (aqueous)	7.5 ml

F. Phosphate Buffered Glutaraldehyde (2.5%)/Formaldehyde (2%)

Working 0.1 *M* Solution	50	ml
10% Paraformaldehyde (aqueous)	20	ml
25% Glutaraldehyde (aqueous)	10	ml

G. Phosphate Buffered Acrolein (10%)

Working 0.1 *M* Solution	50	ml
Acrolein	10	ml

H. Phosphate Buffered Acrolein (1%)/Glutaraldehyde (2.5%)

Working 0.1 *M* Solution	50	ml
Acrolein	1	ml
25% Glutaraldehyde (aqueous)	10	ml

Notes

1. Refrigerate all solutions. When cloudiness or microorganism growth is observed, discard.
2. The same phosphate buffer may be used for preparing both aldehyde and osmium fixatives.

Reference

Gomori, G. (1946). Buffers in the range of 6.4 to 9.6. *Proc. Soc. Exptl. Biol. Med.* **62**:33.

SODIUM PHOSPHATE BUFFER: FORMULATION II (0.135 *M*)

A. Stock Solution: 0.135 *M* phosphate buffer

$NaH_2PO_4 \cdot H_2O$	1.50	gm
$Na_2HPO_4 \cdot 7H_2O$	15.20	gm
Distilled water to make	500	ml

This buffer has pH 7.35, and osmolarity 298 m*OsM.*

B. The Preparation of Buffered Fixatives

This is identical to the dilutions in "Sorenson's Sodium Phosphate Buffer: Formulation I." The only difference is that buffer tonicity is 0.135 *M* in this formulation.

References

Manusbach, A. B. (1966). The influence of different fixatives and fixation methods on the ultrastructure of rat kidney proximal tubule cells. II. Comparison of different perfusion fixation methods and of glutaraldehyde, formalin, and osmium tetroxide fixatives. *J. Ultrastr. Res.* **15**:242. Also see Karlsson, U. and R. L. Schultz (1965). Fixation of the central nervous system for electron microscopy by aldehyde perfusion. I. Preservation with aldehyde perfusates *versus* direct perfusion with osmium tetroxide with special reference to membrane and the extracellular space. *J. Ultrastr. Res.* **12**:160 for additional phosphate buffers.

VERONAL-ACETATE BUFFER AND OSMIUM TETROXIDE

A. Stock Solution: Veronal Acetate Buffer

Sodium veronal (sodium barbital)	0.59 gm
Sodium acetate (crystalline)	0.35 gm
Distilled water to make	20 ml

B. Stock Solution: Ringer's Solution

Sodium chloride	40.25 gm
Potassium chloride	2.10 gm
Calcium chloride	0.90 gm
Distilled water to make	500 ml

C. Working Solution

Stock Buffer Solution	20.0 ml
Ringer's Solution	6.8 ml
Distilled Water	50.0 ml
0.1 *N* HCl	22.0 ml (adjust volume for desired pH)

D. Veronal-Acetate Buffered Osmium Tetroxide (2%)

Working Solution	19.5 ml
4% OsO_4 (aqueous)	10.0 ml

Notes

1. The veronal-acetate stock solution rapidly decomposes and should be prepared just before use.
2. Sodium veronal is a barbituate and should be handled carefully.

Reference

Palade, G. E. (1952). A study of fixation for electron microscopy. *J. Exp. Med.* **95**:285.

FORMALDEHYDE: PREPARATION OF AQUEOUS STOCK SOLUTION (10%)

Paraformaldehyde powder	10	gm
Double distilled water	100	ml

1. Dissolve the powder in water by heating to ∿60°C (working in a fume hood).
2. While warm, add 1.0 *N* NaOH dropwise until the solution clears.
3. Cool before use.
4. Refrigerate in glass-stoppered bottle. If a precipitate forms during storage, discard.

Note

This method ensures that the working solution of formaldehyde is methanol-free.

Author Index

Abermann, R., 134, 135, 136, 144
Abraham, J. L., 151, 160
Abrunhosa, R., 71, 80
Adams, C. M. W., 74, 81
Adelstein, P. Z., 41
Aita, S., 108, 114
Aito, Y., 60, 64, 81
Alberty, R., 96, 105
Albin, T. B., 66, 67, 81
Albrecht, R. M., 48, 52, 70, 81, 110, 111, 112, 113, 118, 119, 121, 123, 125, 128
Aldrich, H. C., 68, 95
Alexa, G., 60, 68, 81
Altman, J. H., 41, 69, 81
Altman, P. L., 69
Amsterdam, A., 74, 81
Anderson, G. S., 139, 140, 146
Anderson, K., 41
Anderson, T. F., 96, 98, 99, 101, 105, 116, 121
Arborgh, P., 55, 56, 81
Ardenne, M. von, 30
Artvinli, S., 65, 81
Ashworth, C. T., 73, 81
Aso, C., 60, 64, 81

Baccetti, B., 123, 128
Baggett, M. C., 41
Bahr, G. F., 74, 75, 81, 86
Baker, F. L., 124, 128
Baker, J. R., 65, 70, 81
Baker, N. V., 124, 128
Ball, R. C., 41
Banfield, G. W., 79, 81
Barbi, N. G., 30
Barer, R., 90
Barlow, D.I., 109, 113
Barnes, C. D., 69, 81
Barrnett, R., 73, 89
Bartlett, A. A., 96, 105, 116, 121
Bassett, L. A., 124, 128
Bayliss, O. B., 74
Becker, R. P., 151, 160
Beer, M., 78, 95
Begnin, J., 136, 145
Bemrick, W. J., 127, 128
Bennet, H. S., 58, 81, 168
Benscome, S. A., 79, 82
Bernhard, W., 97, 106
Berry, R. W., 146
Bertaud, W. S., 41
Bessis, M., 118, 121, 125, 129
Bibel, D. J., 125, 129
Biberfeld, P., 63, 64, 65, 84
Billings-Gagliardi, S., 101, 105, 112, 113
Bistricky, T., 80, 82, 125, 130
Black, J. T., 10, 19, 26, 27, 30
Bodian, D., 63, 72, 82
Bone, Q., 56, 57, 75, 82
Boni, A., 41
Bowes, J. H., 53, 59, 60, 65, 68, 82
Boyde, A., 30, 41, 48, 52, 56, 57, 62, 63, 69, 70, 79, 80, 82, 96, 99, 100, 101, 105, 108, 109, 110, 111, 112, 113, 119, 120, 121, 151, 164
Boyde, S., 99, 100, 105
Bradley, D. E., 133, 135, 144

Brandao, I., 58, 92
Braten, T., 133, 139, 144
Breathnach, A. S., 72, 82
Brightman, M. W., 78, 82
Brocker, W., 151, 160
Broers, A. N., 135, 144
Bronskill, J. F., 124, 129
Brown, J. A., 124, 129
Brown, J. N., 101, 105
Brunk, U., 52, 55, 58, 82
Brunner, M. E. G., 138, 144
Buckley, I. K., 62, 82
Bulger, R. E., 72, 77, 94
Burgess, A., 108, 109, 113
Burkl, W., 75, 82
Burstyn, H. P., 96, 105, 116, 121
Buss, H., 62, 71, 83
Busson-Mabillot, A., 58, 62, 64, 83

Carr, K. E., 125, 129
Carroll, J. S., 41
Carstensen, E. L., 56, 68, 83
Carter, H. W., 153, 160
Case, N. M., 58, 66, 92
Casley-Smith, J. R., 73, 83
Cater, C. W., 53, 59, 60, 65, 68, 82, 83
Caughey, R. C., 152, 163
Caulfield, J. B., 55, 57, 75, 83
Chambers, R. W., 63, 83
Chandler, J. A., 30
Chapman, D., 73, 83
Chatfield, E. J., 126, 129
Chisalita, D., 60, 83
Chopra, K. L., 138, 144
Clark, B., 151, 163
Clark, J. M., 62, 71, 83, 133, 144
Clark-Jones, R., 41
Claude, A., 55, 83
Clayton Jr., J. W., 103, 105
Cohen, A. L., 55, 62, 79, 80, 83, 98, 99, 100, 101, 102, 103, 105, 109, 113, 118, 122
Cohen, W. D., 124, 129
Cole, G. T., 112, 113
Collins, R. J., 73, 83
Collins, V. P., 62, 79, 83
Colvin, J. R., 119, 122
Conger, A., 60, 83
Cosslett, V. E., 90
Costa, J. L., 118, 122
Costello, M. J., 109, 113
Coulter, H. D., 108, 113
Crewe, A. V., 30
Criegee, R., 73, 83, 84
Crosby, P., 108, 115
Csapo, Z., 58, 88
Cylo, V. V., 135, 146
Czarnecki, C. M., 61, 84

Daniels, F., 96, 105
Darley, J. J., 79, 84
Davey, D. F., 56, 84
Dawson, R. M. C., 57, 84
De Bault, L. E., 68, 84, 98, 105
De Bruijn, W. C., 61, 74, 84
de Harven, E., 48, 49, 72, 89, 119, 120, 122, 123, 129, 140, 143, 144
De Jong, D. W., 84
Demierre, G., 79, 87
Demsey, A., 52, 72, 84
De Nee, P. B., 48, 143, 144, 148, 151, 160, 161
Denton, E. J., 56, 57, 82
De Petris, S., 78, 84
Deutsch, K., 63, 65, 84
Dewar, C. L., 125, 129
Diechmann, W. B., 103, 105
Dillon, M. J., 126, 129
Dittmer, D. S., 69, 81
Dole, S., 101, 106
Donaldson, L. A., 31
Doty, P., 66, 86
Draenert, K., 108, 112, 113
Draenert, S., 108, 112, 113
Dreher, K. D., 74, 84
Drier, T. M., 48, 62, 84, 125, 129
Dripps, R. D., 103, 107
Drochmans, P., 56, 57, 64, 85

Ealker, J. F., 65, 84
Echlin, P., 48, 108, 109, 110, 113, 131, 137, 138, 139, 140, 142, 143, 144, 145, 150, 154, 161
Eins, S., 99, 106
Eisenstat, L. F., 70, 84
Elleder, M., 74, 84
Ellis, E. A., 109, 113
Engel, C. E., 41

Epling, G. P., 57, 75, 84
Erasmus, D. A., 30
Ericsson, J. L. E., 57, 62, 63, 64, 65, 72, 84, 94
Erk, I., 75, 95
Estable-Puig, R. F., 155, 161
Etherington, L. D., 69, 81
Everhart, T. E., 30
Eyring, E. J., 66, 84
Ezoe, H., 79, 84

Fahimi, F. D., 56, 57, 64, 85
Falk, R. H., 59, 62, 68, 70, 85, 149, 161
Farnell, G. C., 42
Farquhar, M. G., 78, 85
Fassett, D. W., 103, 105
Faulkin, L. J., 63, 71, 90
Feder, N., 62, 68, 85
Fedorko, M. E., 62, 77, 86
Finke, E. H., 63, 88, 98, 107
Flitney, E. W., 64, 65, 66, 68, 85
Flood, P. R., 56, 85, 137, 138, 145, 146
Florkin, M., 92
Fluck, D. J., 73, 83
Forbes, M., 63, 85
Fotland, R. A., 42
Franke, W. W., 77, 85
Franks, F., 108, 109, 113
Franks, J., 141, 145
Fuchs, H., 109, 114
Fuchs, W., 108, 109, 114
Fujita, T., 30, 151, 156, 161

Gale, J. B., 52, 57, 58, 85
Gallagher, J. E., 125, 129
Geissinger, H. D., 151, 153, 161
Geller, J. D., 141, 145
Gennaro, J. F., 55, 56, 57, 58, 59, 72, 85, 92, 149, 151, 152, 163
Gerarde, H. W., 103, 105
Germinario, L. T., 150, 162
Gershman, H., 106
Gertz, S. D., 70, 85
Gibson, E. D., 4, 31
Gil, K. H., 58, 70, 75, 85
Gillett, R., 64, 85
Gilman, A., 107
Glagov, S., 62, 71, 83
Glang, R., 138, 145, 146, 147
Glassman, L. H., 41
Glauert, A. M., 56, 58, 62, 65, 70, 71, 75, 85
Glider, W. V., 133, 143, 146
Gloster, J. A., 79, 94
Goldman, M. A., 153, 161
Goldstein, J. I., 30
Gomori, G., 58, 85, 170
Gonzalez-Aguilar, F., 65, 86
Good, N. E., 57, 86
Goodman, L. S., 107
Goodman. S. L., 153, 161
Gould, K. E., 71, 86
Gray, P., 92
Greco, J., 66, 95
Greer, R. T., 30
Gresham, G. A., 66, 87
Griffith, W. P., 74, 90
Griffiths, K., 134, 145
Grivet, P., 30
Gunning, B. E. S., 62, 88
Gusnard, D., 101, 106, 112, 114

Habeeb, A. F. S. A., 59, 60, 86
Haggis, G. H., 108, 114
Hagler, H. F., 136, 145
Hagstrom, L., 75, 86
Hake, T., 74, 86
Halioua, M., 42
Hall, D. J., 103, 106
Hall, P. M., 146
Hallowes, R. C., 153, 161
Hamilton, J. F., 41
Hammer, R. F., 127, 128
Hampton, J. C., 86
Harris, M. T., 146
Harris, P. J., 52, 62, 88
Harris, W. S., 103, 106, 135, 145
Hart, R. G., 136, 145
Harvey, D. M. R., 80, 93
Haselkorn, R., 66, 86
Haudenschild, C. C., 71, 86
Hayat, M. A., 32, 41, 42, 48, 49, 53, 56, 61, 67, 73, 76, 78, 86, 88, 105, 113, 114, 115, 121, 122, 128, 129, 130, 144, 145, 146, 161, 162, 163
Hayek, K., 134, 145
Hayes, T. L., 30, 48, 110, 114, 123, 129
Hayunga, E. G., 124, 129

Henderson, W. J., 134, 145
Heslop-Harrison, J., 149, 161
Heslop-Harrison, Y., 149, 161
Hesse, I., 119, 122
Hesse, W., 119, 122
Hicks, R. M., 65, 86
Higgins, G. C., 42
Hillman, H., 63, 65, 84
Hillson, P. J., 42
Hintermann, H. E., 136, 145
Hiramoto, R., 59, 60, 86
Hirsch, J. G., 62, 77, 86
Hobbs, M. J., 75, 86
Hodges, G. M., 153, 161
Hohn, F. T., 138, 145
Honjou, K., 136, 145
Holland, L., 134, 145
Holland, V. F., 139, 143, 145
Holt, D. B., 30
Holt, S. J., 65, 86
Hopwood, D., 53, 55, 56, 60, 61, 64, 65, 66, 68, 74, 77, 86, 87
Horisberger, M., 153, 161
Horl, E. M., 109, 114
Howard, K. S., 48, 79, 87
Howden, H. F., 149, 161
Howell, P. G. T., 41, 42
Hoyer, L. C., 153, 161
Hren, J. J., 30
Humphreys, W. J., 48, 53, 87, 101, 103, 106, 108, 110, 111, 114
Huxley, H. E., 61, 78, 80, 87, 90
Hyde, P. J. W., 140, 145
Hyzer, W. G., 42

Idle, D., 152, 162
Igbal, S. J., 55, 57, 63, 64, 87
Ingram, P., 49, 119, 120, 121, 122, 140, 143, 145
Irino, S., 151, 156, 162
Irish, D. D., 106
Itoshima, T., 151, 156, 162
Iwata, H., 108, 114

Jacks, T. J., 80, 87, 90
Jacobson, C. I., 42
Jacobson, R. E., 42
Jacques, W. E., 152, 162
Jalanti, T., 79, 87
James, T. H., 42
Jansen, E. F., 59, 61, 87
Jard, S., 56, 87
Jenkins, R. L., 42
Johari, O., 48
Johnson Jr., J. E., 62, 87
Johnston, P. V., 62, 80, 87
Johnston, W. H., 63, 70, 87
Jones, A. L., 151, 153, 163
Jones, A. V., 31
Jones, D., 61, 66, 87
Jones, G. J., 64, 87
Jones, J. B., 154, 159, 162
Jordan, F. I., 42
Jost, P., 56, 88
Joy, D. C., 31

Kahn, L. E., 79, 80, 88
Kalimo, H., 70, 71, 88
Kallman, F., 74, 91
Kalt, M. R., 62, 68, 88
Kamler, H., 151, 153, 161
Kanaya, K., 136, 145
Karlsson, U., 62, 64, 70, 71, 88, 92, 171
Karnovsky, M. J., 56, 68, 71, 88, 92
Kay, D. H., 144
Kaye, G., 154, 161
Keino, H., 66, 92
Kellenberger, E. A., 61, 77, 78, 88, 92, 111, 114, 168
Kelley, R. O., 152, 154, 155, 156, 162
Kenten, R. H., 65, 82
Kenway, P. B., 41
Kerr, T. J., 125, 126, 130
Kessel, R. G., 31
Khattab, F. I., 70, 94
Kiernan, J. A., 77, 88
Kirillou, N. J., 42
Kirkpatrick, K., 40, 42
Kirschner, R. H., 101, 106, 112, 114
Kirz, J., 31
Kistler, J., 111, 114
Klainer, A. S., 48, 119, 122, 125, 129
Knowles, J. R., 59, 60, 91
Koch, G. R., 108, 114, 150, 162
Koller, T., 97, 106
Kormendy, A. C., 49, 62, 88, 125, 127, 128, 129
Kotelianski, V. E., 136, 147

Kuhn, C., 63, 88
Kunoh, H., 151, 156, 162
Kuran, H., 57, 58, 88
Kurtzman, C. P., 103, 106
Kushida, T., 151, 162
Kuthy, E., 58, 88

Lamb IV, J. C., 49, 119, 120, 121, 122
Landis, W. J., 66, 88
Langenberg, W. G., 63, 88
Langford, M. J., 42
Larsson, L., 63, 70, 88
Laudate, A., 151, 163
Lawson, J. W., 125, 129
Lawton, J., 52, 62, 88
Le B Skaer, H., 109, 114
Ledbetter, M. C., 62, 88, 108, 114, 149, 150, 152, 154, 155, 162, 164
Lee, R. M. K. W., 55, 57, 62, 78, 88, 100, 101, 106
Le Furgey, A., 123, 129
Leif, R. C., 119, 122, 153, 161
Leister, D. A., 41
Leppard, G. G., 119, 122
Levy, W. A., 61, 88
Lewis, E. R., 97, 98, 100, 106
Liepins, A., 49, 73, 89, 119, 120, 122
Lin, C. H., 80, 89, 99, 106
Ling, L. E. C., 149, 161
Litman, R. B., 73, 89
Lojda, A., 66, 89
Lojda, Z., 74, 84
Lombardi, L., 78, 89
Ludwig, H., 31
Luft, J. H., 57, 58, 62, 66, 74, 81, 89, 94, 95, 168
Lung, B., 123, 129

Macconachie, E., 52, 56, 57, 62, 79, 80, 82, 99, 101, 105, 112, 113
Machado, A. B. M., 72, 89
MacKenzie, A. P., 48, 108, 110, 111, 112, 113, 114, 119, 121, 123, 128
Mahr, S. C., 73, 93
Maissel, L. I., 140, 145, 146, 147
Malhotra, S. K., 55, 89
Malick, L. E., 152, 154, 155, 156, 162
Mangan, J. L., 59, 62, 94
Marchant, H. J., 49, 118, 122, 123, 124, 129
Marinozzi, V., 57, 61, 72, 73, 74, 90
Marshall, A. T., 109, 112, 114
Martin, M., 72, 82
Maruszewski, C. M., 31
Maser, M. D., 57, 89, 99, 106
Maser, M. M., 80, 89
Mason, H. S., 92
Matthieu, O., 56, 57, 63, 64, 71, 89
Maugel, T. K., 49, 58, 63, 89, 117, 122, 124, 130
Maunsbach, A. B., 57, 62, 69, 70, 71, 89, 171
Maupin-Szamier, P., 75, 89
McAlear, J. H., 150, 162
McCrae, J. M., 65, 81
McCully, M. E., 52, 53, 62, 63, 65, 66, 68, 70, 74, 89
McDowell, E. M., 63, 65, 68, 89
McGee-Russell, S. M., 42
McKee, A., 125, 128, 130
McKinley, R. G., 57, 89
Mees, C. E. K., 42
Mellor, S., 101, 106
Mersey, B., 52, 53, 62, 63, 65, 66, 68, 70, 74, 89
Mesulam, M. M., 91
Metzger, H., 31
Millonig, G., 57, 58, 61, 72, 73, 74, 75, 80, 89, 90
Mittler, B. S., 61, 91
Molcik, M., 135, 145
Molday, R. S., 153, 162
Mollenhauer, H. H., 90
Moncur, M. W., 101, 106
Moore, D. J., 90
Morel, F. M. M., 62, 90
Moreton, R., 150, 161
Moretz, R. C., 59, 65, 74, 90
Moore, D. E.
Moss, G. I., 56, 62, 90
Mowczko, W. E., 59, 76, 93
Mozingo, H., 149, 162
Muir, M. D., 31
Muller, L. L., 80, 87, 90
Mullins, J. T., 109, 113
Mumaw, V. R., 78, 90, 152, 154, 155, 156, 163
Munawar, M., 125, 130

Munger, B. L., 49, 78, 90, 131, 143, 146, 148, 152, 154, 155, 156, 163
Murakami, T., 151, 152, 153, 156, 158, 159, 163
Murphy, J. A., 148, 150, 152, 153, 155, 157, 158, 160, 163

Nagatani, T., 134, 146
Nagy, I. Z., 109, 112, 114
Neblette, C. B., 42
Nei, T., 108, 109, 114, 149, 150, 163
Nemanic, M., 42, 110, 115
Nermut, M. V., 112, 115
Neugebauer, C. A., 134, 146
Newburg, D. E., 31
Newell, D. G., 124, 125, 130
Newman, H. N., 110, 115
Nickerson, A. W., 117, 122, 123, 130
Niedrig, H., 138, 146
Nielson, A. J., 74, 90
Nixon, W. C., 31
Nordlund, U., 57, 62, 63, 91
Norton, T. N., 66, 90
Nowell, J. A., 52, 55, 59, 62, 63, 68, 69, 71, 79, 90, 152, 154, 163

Oatley, C., 31
O'Brien, T. P., 52, 55, 62, 68, 85, 90
Ockleford, C. D., 72, 90
Ofengand, J., 66, 84
Ogura, K., 151, 163
Okagi, T., 151, 163
Olzewski, M. J., 57, 58, 88
Oster, G., 93, 105

Page, S. G., 80, 90
Palade, G. E., 55, 58, 78, 85, 90, 172
Palay, S. L., 70, 90
Panayi, D. N., 139, 140, 143, 146
Panessa, B. J., 149, 151, 152, 163
Parsons, D. F., 81, 91
Parsons, E., 62, 80, 101, 106, 118, 122
Pawley, J. B., 48, 52, 55, 59, 62, 63, 68, 69, 71, 79, 90, 101, 106, 110, 114, 123, 129
Pearson, S., 138, 146
Pease, D. C., 110, 115
Pederson, M., 124, 130
Pendergrass, R. E., 124, 128
Pentilla, A., 52, 64, 74, 77, 91, 101, 106
Perrachia, C., 61, 91
Peters, K. R., 138, 146, 154, 163
Pexieder, T., 57, 91
Pfefferkorn, G., 31, 49, 151, 160, 163
Pillard, T. D., 75, 89
Pilstrom, L., 57, 62, 63, 91
Pladellorens, M., 62, 65, 91
Pliskin, W. A., 138, 146
Polliack, A., 118, 122
Pollister, A. W., 93, 105
Porter, K. R., 74, 91
Porter, M. C., 126, 130
Postek, M. T., 48, 79, 80, 87, 91, 118, 122, 152, 155, 163
Prento, P., 80, 91
Princen, L. H., 124, 128
Priyokumas-Singh, S., 138, 146

Quiocho, F. A., 59, 91

Raistrick, A. H., 82
Rajaraman, R., 62, 95
Ramirez-Mitchell, R., 112, 113
Rebhun, L. I., 108, 110, 112, 115
Reese, T. S., 78, 82
Reimer, L., 31
Revel, J. P., 31
Rhoades, K. R., 125, 129
Rice, R. M., 103, 107
Richards, F. M., 59, 60, 91
Riemersma, J. C., 73, 91
Roath, S., 124, 130
Robards, A. W., 108, 115
Robertson, E. A., 61, 91
Robertson, J. G., 61, 91
Rochow, E. G., 31
Rochow, T. G., 31
Rodman, N. F., 152, 163
Roli, J., 138, 146
Roomans, G. M., 31
Roots, B. I., 62, 80, 87
Rosario, B., 65, 95
Roschger, P., 109, 114
Rosenberg, W., 103, 107
Rosene, D. C., 91
Rosen, J., 106
Rosenkrantz, H., 70, 81
Rosset, J., 153, 161

Rowsowski, J. R., 133, 143, 146
Rudman, R., 58, 91
Ryan, K. P., 56, 57, 75, 82
Ryter, A., 77, 92, 168

Sabatini, D. D., 58, 59, 65, 66, 68, 72, 92, 167
Saito, M., 66, 92, 134, 146
Salema, R., 58, 92
Salpeter, M. M., 135, 136, 144
Sanders, S. K., 123, 125, 130
Saubermann, A. J., 48, 108, 109, 113, 115
Sax, N. I., 76, 92
Schidlovsky, G., 73, 92
Schiechl, H., 72, 75, 82, 92
Schiff, R., 55, 56, 57, 58, 59, 72, 92
Schlatter, C., 77, 92
Schlatter-Lanz, I., 77, 92
Schmalbruch, H., 53, 63, 69, 92
Schneeberger-Keeley, E. E., 56, 68, 71, 92
Schneider, G. B., 49, 101, 107, 108, 112, 115
Schramm, M., 74, 81
Schultz, R. L., 58, 61, 62, 64, 66, 70, 71, 88, 91, 92, 171
Schur, K., 138, 146
Schwabe, U., 134, 145
Scott, J. R., 103, 107
Sechaud, J., 61, 78, 92
Seiler, H., 31
Sela, A., 151, 164
Seligman, A. M., 152, 154, 164
Shapiro, B. L., 152, 157, 158, 164
Shaw, J., 56, 92
Shay, J. W., 59, 62, 93
Shaykr, M. M., 62, 83, 101, 105
Shelton, E., 59, 76, 93
Sherman, D. M., 140, 146
Shiflett, C. C., 132, 134, 146
Shih, C. Y., 31
Shved, A. D., 135, 146
Siew, S., 154, 164
Silva, M. T., 61, 78, 93
Silverberg, B. A., 80, 82
Silverglade, A., 103, 107
Simionescu, M., 152, 164
Simionescu, N., 152, 164
Simmens, S. C., 143, 146
Simon J. A. V., 52, 93
Sirak, H. D., 62, 93
Sjostrand, F. S., 57, 65, 75, 84, 93
Skaer, R. J., 62, 93
Slayter, H. S., 133, 134, 135, 136, 137, 146
Sleigh, M. A., 109, 113
Sleytr, U. B., 108, 115
Slobodrian, M. L., 142, 147
Small, E. B., 49, 124, 130
Smith, C. W., 81
Smith, M. E., 98, 107
Sogard, M., 151, 160
Speck, R., 42
Speizer, F. E., 103, 107
Sperelakis, N., 63, 85
Spicer, R. A., 110, 115
Spiller, E., 135, 144
Spurr, A. R., 31
Stein, O., 73, 75, 93
Stein, Y., 73, 75, 93
Stoeckinius, W., 73, 93
Stoner, C. D., 62, 93
Stroke, G. W., 42
Stumpf, W. E., 63, 71, 93
Subirana, J. A., 62, 65, 91
Sweney, C. R., 124, 130, 152, 157, 158, 164

Tahmisian, T. N., 57, 93
Tailby, P. W., 110, 115
Takahashi, G., 152, 158, 164
Tandler, B., 62, 68, 88
Tannenbaum, M., 151, 153, 164
Tarrant, P., 105
Teetsov, A., 124, 129
Terracio, L., 108, 113
Terzakis, J. A., 78, 93
Thornley, M. J., 62, 85
Thornwaite, J. T., 55, 57, 62, 93, 119, 122
Thorpe, J. R., 80, 93
Thurston, E. L., 48, 53, 62, 66, 84, 93, 125, 129
Thurston, R. Y., 75, 93
Ting-Beall, H. P., 78, 94
Tobin, T. P., 56, 57, 94
Todd, R. L., 125, 126, 130
Tokunaga, J., 151, 156, 164
Tombes, A. S., 123, 130
Tomimatsu, Y., 61, 94
Toner, P. G., 125, 129

Tooze, J., 55, 94
Tormey, J., 74, 79, 94
Tovey, N. K., 42
Trimble, J. J., 80, 89, 99, 106
Trnavska, Z., 68, 94
Troughton, J., 31
Trump, B. F., 57, 63, 65, 68, 72, 77, 89, 94
Tsutsumi, V., 79, 82
Tucker, S. C., 80, 91, 152, 155, 163
Tyler, W. S., 68, 94

Umrath, W., 110, 115
Unitt, B. M., 30
Urbach, F., 42
Usherwood, P., 57, 89

Valentine, R. C., 78, 94
van Deurs, B., 62, 94
van Duijn, P., 55, 66, 94
Van Harreveld, A., 70, 94
Vasilier, V. D., 136, 147
Verhoeven, J. D., 4, 31
Vesely, P., 56, 62, 82, 119, 120, 121
Vogel, G., 153, 164

Wadsworth, N. J., 138, 146
Walker, C., 59, 62, 93
Walker, E. R., 143, 144, 148, 161
Wall, E. J., 42
Ward, B. J., 79, 94
Waterman, R. E., 62, 71, 94, 118, 122, 154, 164
Watson, L. P., 59, 62, 63, 94, 117, 122, 125, 126, 127, 128, 130
Wayman, M., 125, 129
Weakley, B. S., 55, 57, 58, 63, 64, 87, 94
Weed, R. I., 118, 121, 125, 129
Wehner, G. K., 139, 140, 147
Weibel, E. R., 58, 70, 75, 85
Wells, O. C., 32, 48
Welter, L. M., 4, 32
West, J., 59, 62, 94
Wetzel, B., 52, 70, 81, 94
Wheeler, E. E., 109, 115
White, D. L., 73, 95
Whytock, S., 62, 93
Wilhelms, E., 99, 106
Williams, R. C., 132, 147
Williams, S. T., 62, 95, 128, 130
Willison, J. H. M., 62, 95
Wilson, D. C., 153, 159, 164
Wilson, R. B., 152, 154, 155, 156, 162
Wold, F., 59, 95
Wolfe, S. L., 78, 95
Wollman, H., 66, 95, 103, 107
Wolman, M., 65, 66
Wood, C., 109, 110, 111, 113
Wood, R. L., 57, 58, 74, 89, 95
Woods, P. S., 150, 152, 154, 155, 164
Wrigglesworth, J. M., 53, 56, 95
Wyckoff, W. G., 132, 147

Yakowitz, H., 30
Yamamoto, I., 65, 95
Yoshii, Z., 32
Young, J., 32

Zalokar, M., 75, 95
Zanin, S. J., 138, 146
Zeikus, J. A., 68, 95
Zingsheim, H. P., 133, 147
Zobel, C. R., 78, 95
Zubay, G., 61, 78, 87

Subject Index

Accelerating voltage, 5, 10, 11, 149
Acetone
  air-drying, 118
  dehydration, 79, 98, 100
  extraction, 79, 80
  infiltration fluid, 98, 100
  surface tension, 118
Acrolein, 54, 66, 67
  extraction, 66
  fixation rate, 66, 68
  mixtures, 66, 68, 69, 70
  penetration rate, 66, 68
  toxicity, 66, 67
Air-drying, 46, 116, 117
  acetone, from, 118
  anhydrous, 119
  argon, with, 119, 120
  comparison with CPD, 119, 120, 121
  ethanol, from, 119, 120
  Freon, from, 118, 119, 120
  results of, 120, 121
  vacuum, 119
  water, from, 98, 116, 118, 120
Alcohol. *See* specific reagents
Aldehydes
  comparison of, 54, 67, 68, 69
  concentration, 57
  mixtures, 54, 66, 68, 69, 70, 71
  storage solution, 65, 68
  tonicity, 56, 57
  also see specific aldehydes
Aldol condensation, 60
Alignment, 8
Aluminum, thin films, 136
Amino acids, 53, 60, 74
Amyl acetate
  freeze-drying from, 109
  infiltration with, 99, 100
  intermediate fluid, 98, 99, 103
  toxicity, 103
Anaesthetics, 69, 124
Animal tissues
  dehydration, 79, 80
  fixation, 62, 68, 69
  also see specific tissues
Amplifier, video, *2*
Anode, *2*, 4
Anoxia, 70
Apertures, *2*, 5, 6, 8, 25, 38
  alignment, 8
  cleaning, 8
Aquatic organisms, 58, 117, 124
  also see specific organisms
Argon
  air-drying, 119, 120
  sputtering, 140, 141
Artifacts. *See* Shrinkage
Astigmatism, 8, 25, 26
Auger electrons, 9, 10
Autolysis, 53, 56, 63, 69

Backscattered electron detector, 9, *10*, 12
Backscattered electrons, 8, 9, *10*, 47, 48, 131, 138, 150

Bacteria, 43
  air-drying, 118, 119
  fixation, 62, 77, 128, 168
  handling, 124, 125
Baskets
  for CPD, 101, 103, 104
  for evaporation, 132, 133
  for handling small organisms, 124
Beam diameter. *See* Spot size
Bell jar
  apparatus, 132
  maintenance, 139
Bias, *2*, 4
Bladder, 119, 120
Blood
  cells, 62, 80, 118, 119, 120, 124, 125
  vessels, 62, 71
Bone, 66, 112, 151
Brain, 62, 71, 80, 99
Buffers, 57, 58, 59
  ionic composition of, 55, 57
  natural, 55, 58
  purpose, 53, 55, 56
  wash, 72
  also see specific buffers

Carbohydrates, 66, 74
Carbon dioxide
  characteristics, 98, 100
  transitional fluid, 98, 99, 104
Carbon evaporation, 133, 136
  method, 136, 137
  platinum, and, 135
Carbonyl compounds, 55
Cartilage, 108, 112, 119
Cathode ray tube, *2*, 5, 12, 13, 14, 33, 34
Cathodoluminescence, 1, *10*, 12, 109, 136, 151
  detector, 12
Cations
  divalent, 57, 58
  monovalent, 57, 58
  also see Electrolytes
Central nervous system, 70
Centrifugation, 123, 124
Charging, 123, 124
Chloroform, cryoprotectant, 109
Chromatic aberration, 23, 24
Collagen, 65, 68
Compression, 44
Condenser lens, 3
Conductive paint
  carbon, 46
  silver, 46, 149
Contamination, 13, 14, 19, 29
Contrast, 5, 11, 37, 39
Critical opalescence, 97
Critical point, 97
Critical point drying, 46, 96–104
  apparatus, 101, *102*
  comparison with FD, 112
  hazards, 101, 102, 103
  method, 104
  operation, 103
  theory, 96, 97
Critical pressure, 97
Critical temperature, 97
Cryoprotectants, 109, 110
Cultured cells and tissues
  air drying, 119
  dehydration of, 79, 100
  fixation of, 62

Deflection coils, *2*, 7, 13
Dehydration
  chemical, 50, 79, 80
  extraction during, 79, 80
  see also specific reagents and drying methods
Demagnifying lenses, *2*, 5, 7
Density, atomic, 9, 10
Detection system, 12, 13
Detective quantum efficiency, 36
Detectors, signal, *2*, 3, 9, *10*, 12, 13
  also see specific detectors
Diatoms, 51, 118
Diffraction, 20–23
Diffusion pump
  bell jar, 132
  SEM, *16*, 17, 18, 19
Dimensional changes, 52
  also see Shrinkage
2,2-dimethoxypropane, 79, 80, 98, 99, 100
Dimethylsulfoxide, 109
Direct current sputtering, 140, *141*
Display system, *2*, 3, 13
Double fixation, 52

Elastic collisions, 9, *10*
Electrolytes, 56, 57
Electron gun, *3*, 4, 5, 6
  cleaning, 6, 7
Electron signals, 9
  also see specific signals
Embryonic tissues, 62, 68, 71, 118
Emulsions, 34
Endoplasmic reticulum, 62
Energy-dispersive X-ray analysis, 1, *10*, 12, 126, 136, 138, 151, 153
Ethanol
  air-drying, 119, 120
  dehydration, 79, 80, 98, 99
  infiltration fluid, 98, 100, 104
  surface tension, 118
Exposure, 34
Extraction, 52, 66, 73, 77, 109
  during dehydration, 73, 79, 80, 101
  during fixation, 55, 66, 72, 73, 77, 101
  also see Shrinkage
Evaporated coatings, 23, 47, 48, 50, 131–139
  apparatus, 132
  artifacts, 133, 148
  comparison with metallic impregnation, 160
  comparison with sputtering, 143
  criteria, 135
  mass thickness, 137
  method, 133, 134, 139
  resolution, 137
  theory, 131, 134
  thickness, 133, 135, 137, 138
  also see Carbon evaporation and specific metals

Fatty acids, 61, 66
Field emission guns, 4, 15, 28
Filaments, *2*, 3, 4, 6, 14
  exchange of, 6
  saturation, 4, 6
Film, 38, 39
Filtration, 118, 123
  applications, 125
  filter types, 125, 126
Fixation
  chemical, 50, 52
  duration, 55
Fixation (*cont.*)
  penetration rate, 53, 64, 67, 68
  primary, 53–55
  rate of, 64, 67, 68
  secondary, 55, 72–77
  temperature, 63, 64
  tertiary, 52, 61, 77, 78
  also see specific fixatives
Fixation methods, 53, 67, 69–71
  immersion, 59, 69, 70
  in vivo, 69
  microinjection, 69, 71
  vascular perfusion, 63, 69, 70, 71, 158
Fluorocarbons. *see* Freon
Focus, 7, 8
Foraminifera, 117, 123
Formaldehyde, 54, 65, 66, 172
  fixation rate, 68
  mixed with glutaraldehyde, 65, 68, 70, 71
  penetration rate, 65
  preservation, 65
  reaction, 65, 66
Free-living cells, 123–128
  also see specific cell type
Freeze-drying, 45, 108–112
  apparatus, 110
  comparison with CPD, 112
  method, 110, 111
  rate, 111
  shrinkage, 112
  theory, 112
Freeze fracturing, 108, 109, 110, 111, 112, 155
Freeze substitution, 109
Freezing point depression, 57
Freon
  air drying from, 118, 119–120
  TF, 99, 100, 104
  13, 98, 100
  surface tension, 118
  toxicity, 103
  transitional fluid, 98
Frozen specimens, 148–150
  examination of, 108, 149
  preparation of, 149, 150

Galloylglucose. *See* Tannin
Gamma, 11, 37, 38

Geometry, sample, 11
  see Tilt
Glutaraldehyde, 54, 59–65
  concentration, 63, 64
  crosslinking, 59, 60, 61, 64
  duration, 63
  effects on ultrastructure, 61, 62
  fixation rate, 64, 68
  handling, 64, 65
  mixture of aldehydes, 65, 66, 68, 69, 70, 71, 77
  mixture, with osmium, 77
  penetration rate, 64, 68
  pH, 63, 64
  reactions, 60, 61
  stock solutions, 64
  temperature, 63, 64
  tonicity, 64
Glycerin, 149
Glycogen, 61, 74
Gold
  characteristics, 135, 136
  evaporated films, 135
  and palladium, 136
  site selective marker, 153
  sputtered films, 142
  thin film removal, 151
Graininess, 36
Grid cap; see Shield
GTA-O-TA-O method, 152, 157

Ice, 109, 112
Ideal gas law, 97
Illuminating system, 1, *3*
Image recording system, 1, 13, 14
Impregnation, metallic, 47, 148, 150–160
  applications, 150–151
  see specific methods
Inelastic collisions, 9, *10*
Information system, 1, 9, *10*
Insects, 117, 149, 157
Interfacial tension, 43, 44, 45, 46, 116
Intermediate fluids, 98, 99, 100
  see specific fluids
Ion beam sputtering, 141
Ionization gauge, 19

Karnovsky's fixative, 68, 71, 120
Kellenberger's fixative, 168
Kidney, 63, 70, 71, 109, 119, 120, 148, 151, 159

Lanthanum hexaboride guns, 4, 14, 28
Latent image, 34
Lead acetate, 152
Ligands, 73, 77, 151
Lipids
  saturated, 61, 73, 74, 80
  unsaturated, 61, 73
Lipoproteins, 74
Liver, 63, 64, 101, 151
Low temp. vacuum drying, 119, 120
Lung, 63, 71

Magnification, 7, 8, 12
Membranes
  fixation of, 62, 74, 78
  permeability of, 56, 74, 75
Metals
  non-refractory, 132, 135
  refractory, 135
Microdissection, 47, 151, 156
Micropipetting, 123–124
Mitochondria, 62
Muscle, 63, 80

Negatives, 33, 39, 40
Nematodes, 123
Nitrous oxide, 97, 98
Noise, electron, 12, 14, 36
Non-electrolytes, 56, 57
Nuclear pore filters, 47, 118, 126, 127
Nucleic acids, 61, 74, 78
Nucleus, 62, 78

Objective lens, 7
Observation screen, 13
O-GTA-O-GTA-O method, 152, 158
Osmium tetramethylenediamine, 153, 159
Osmium tetroxide
  buffered, 75
  concentration, 75
  handling, 76
  ligand-mediated binding, 152–156

mixture, glutaraldehyde, 77
post-fixation, 52, 55, 72–77
preparation, 76, 77
rate of penetration, 75
reactions, 73, 74
regeneration, 77
specimen size, 75
stain effect, 73
temperature, 75
tonicity, 75
toxicity, 76
unbuffered, 55, 75
Osmium tetroxide-α-naphthylamine, 74
OsTMEDA, 159–160
Osmolarity, 55, 56, 72
also see Tonicity
Osmometer, 57
OTO, 152, 154
OTOTO, 152, 154–156
cryofracture, 155
cryoresinfracture, 155
method, 154, 155
Ovary, 63

Paint, conductive, 149
Palladium
characteristics, 135, 136
evaporated films, 135
Paraformaldehyde; see Formaldehyde
pH, 53, 55
animal, 56
physiological, 53, 56, 57
plants, 56
Phosphate buffers, 58, 75, 78, 169–171
Phospholipids, 61
Photography, 7, 8
Photomultiplier, 12, 13
Phytoplankton, 118
Pipes buffer, 58, 59
Pirani gauge, 19
Plant tissues
air-drying, 118
critical point drying, 99
dehydration, 79, 80, 99
fixation, 62, 63, 66, 69, 70
freeze-drying, 108, 109
fresh frozen, 51, 149, 150
metallic impregnation, 151, 152, 155, 157

Plasma sputtering. *See* Sputter coating
Platinum
characteristics, 135, 136
evaporated films, 135
site-selective markers, 153
Pollen, 117
Potassium iodide, 152
Potassium permanganate, 152
Preparation, flow chart of specimen, 51
Preservation, criteria, 50, 52, 67
Printing, 40
Proteins, fixation of, 52, 53, 55, 59, 65, 66, 74

Quenching media, 110
Quenching unit, 110

Radio-frequency sputtering, 140
Radiolarians, 117
Raster pattern, 7
Resistance heating, 131, 134, 136 see Evaporated coatings
Resolution, 1, 5, 8, 13, 14, 19–35, 39, 47, 131, 135
Rotary evaporation, 132, 133, 137, see Evaporated coatings
Rotary pumps, *15*, 16, 17

Safety, 53. *also see* specific chemicals and techniques
Saline, preperfusion washing, 71
Saturation, filament, 4, 6
Scan generator, *2*, 7, 13
Scan rate, 7, 34, 36, 38
Scanning electron microscope, 1, *2*
Schiff's reaction, 55
Scintillator, 10, 13
Secondary electrons, 9, *10*, 131
detectors, 9, *10*, 12
Shield, *2*, *3*, 4, 6
Shrinkage
during air-drying, 118, 120, 121
during CPD, 99, 100, 101, 112, 153
during dehydration, 55, 79, 80, 81, 99, 101, 112, 153
during FD, 109, 112
during fixation, 55, 57, 66, 69, 72, 101

Signal-to-noise ratio
  film, 35
  radiation, 36–38
Silver, site selective marker, 153
Silver halide, 34
Small intestine, 70, 71
Sodium bicarbonate, 58
Sodium cacodylate, 58, 71, 72, 78, 165–166
Sodium phosphate. *see* Phosphate buffers
Spherical aberration, 25
Spleen, 151
Spot size, 5, 8, 26, 27, 28, 149
Spread function, 35
Sputter coating, 23, 47, 48, 50, 127, 136, 139–143
  apparatus, *141*
  artifacts, 143, 148
  characteristics, 142
  comparison with evaporation, 143
  comparison with metallic impregnation, 160
  film thickness, 142, 143
  metals, 142
  method, 141, 142
  plasmas, 139, 140
  theory, 139
  thermal damage, 139, 140, 141
Stigmator, *2*, 8, 26
Substrates
  aluminum, 46, 125
  carbon, 46, 126
  coverslips, 125
Surface tension, 46, 96, 118
Suspensions, cell
  centrifigation, 124, 127
  cleaning, 127
  filtration, 118, 123, 127
sym-Collidine, 58, 59, 75, 167, 168

Tannic acid, 152, 157
Tannin, 65, 73, 152, 156–159
TAO, 152, 153, 156
TAOTO, 153, 156, 157, 159
Tape, 46, 47, 149
Tertiary fixation, 52, 77
Testis, 63
Thiocarbohydrazide, 73, 77, 152, 154, 156
Tilt, effect of, 11
Tonicity, 53, 56, 57, 64, 69, 71
  also see Osmolarity
Trachea, 119, 120
Transitional fluid, 78, 96, 97, 98
  also see specific fluid
Triode sputtering, 140
Tungsten, evaporation, 135, 136

Uranyl acetate, 77, 78, 79
  reactions, 77, 78
  stain effect, 77, 78, 152, 155
  tertiary fixation, 52, 77–79
  use of, 78, 79
Uranyl nitrate, 152
Uterus, 119, 120

Vacuum system, 3, 4, 14–19
Veronal acetate, 58, 171–172
Vibrations, 29
Viruses, 43, 48, 63, 78
Volume-to-surface ratio, 43, 116

Wavelength, 20
Wehneldt cylinder. *See* Shield
Working distance, 11, 12

X-ray analysis, 78, 108, 136

Yeast, 109

THE MASS
FOR CATHOLIC CHILDREN

By

Dr. Kelly Bowring, S.T.D.

&

Rev. Michael J. Sullivan

Illustrations by: Larry Ruppert

Manufactured in Hong Kong through InterPress Ltd.
December 2012 --- Job# 121027

PREPARATION FOR MASS

The Mass is always new because the living Jesus comes to meet us at every Mass. We will find Mass interesting and exciting when we participate fully. Let's go to Mass desiring to meet with Jesus and expecting to be filled with His love and joy, and with Jesus Himself. We must go to Mass at least every Sunday. We are invited to receive Jesus Christ in the Eucharist at every Mass, and we are required to receive Holy Communion at least once a year.

We should make sure we have the proper disposition before receiving Jesus in the Eucharist because He is all good and all holy. We must fast one hour before receiving Holy Communion.

If we have committed a mortal (serious) sin since our last Confession, we must not receive Holy Communion without first receiving absolution in the Sacrament of Penance. So, if we need to, we can go to the priest for Confession before Mass. It is good to say some prayers before and after Mass too.

As Mass begins, we should pay attention and participate by singing the hymns and saying the responses out loud. Ask Jesus to help you and He will.

GOING TO MASS ON SUNDAY

All Catholic Christians are required to attend Holy Mass every Saturday evening or Sunday. The third commandment "Keep holy the Sabbath" must be followed strictly. If possible, it is good to attend Mass together as a family. A family that prays together is a family that stays close to God. God gives us the gift of the Eucharist at Mass.

When we come to Mass on Sunday, we offer to God all that we are and all that we need. Our Lord hears our prayers. We worship God on Sunday and receive the gift of his love.

BLESSING OURSELVES WITH HOLY WATER

Upon entering a church, there is normally a holy water container or font. The font or container contains blessed water, or holy water. The water was blessed by a priest either at a baptism or before Mass. As we approach the font or holy water container we dip our finger (or fingers) in holy water as a reminder of our baptism. Then, we take our finger with holy water on it and make the Sign of the Cross as when we were baptized and the priest poured water over us and said "I baptize you in the name of the Father, Son and Holy Spirit."

GENUFLECTION

When we enter church, we are to genuflect in the direction of the tabernacle. We do this because in the tabernacle Jesus lives in the consecrated bread. It is really and truly Jesus and we genuflect to show respect and love for Him. God deserves nothing less. Jesus is King of heaven and earth. He is our Redeemer and Savior and worthy of all respect. As we reverently genuflect, we may make the Sign of the Cross to show our belief in the Father, Son and Holy Spirit.

BEFORE HOLY MASS

Before Holy Mass, the faithful are called to prepare. We prepare to meet our Lord in the Blessed Sacrament of the Eucharist. Many prepare for Mass by offering prayers such as the rosary or other formal prayers. Others prepare for Mass by praying silently in the pew asking for what they need or giving thanks for what they have.

PROCESSION

Before Mass begins, servers, lectors, other ministers of the Mass, and the priest, line up for the entrance procession. The procession is very important to the Mass. The priest comes last symbolizing our Lord as he entered Jerusalem to celebrate the Last Supper with his disciples. The opening hymn is then played and the procession begins. The entrance procession moves slowly and reverently down the aisle up to the sanctuary and then all take their places. All depart the same way they entered.

THE SIGN OF THE CROSS

As Mass begins, the priest processes to the altar. He bows to kiss the altar, which signifies Christ.

At the chair, he leads us in making the Sign of the Cross. While we call on God, the Father, Son, and Holy Spirit, we are also reminded that Jesus died on the Cross to save us from our sins. The priest leads us in asking God's forgiveness and mercy for our sins, so that we can worthily participate in Mass.

THE OPENING PRAYER

Joining together into one family - God's Family- the priest prays the Opening Prayer and the other Mass prayers aloud while we join him in the silence of our hearts. The whole Mass is a prayer, the greatest prayer of the Church. We answer, "AMEN", which means we believe and agree with the prayer.

THE READINGS

God's Word is found in the Holy Bible, and He is it's Author. The Bible is also called the Sacred Scriptures, and it contains the Old and the New Testament. The Bible is organized for Mass into a Lectionary. All of its main parts are read at Mass over a three-year period. God is present and speaks to us in His Word. At Mass, one or two readings and a Psalm are read before the Gospel. We should listen to the readings at Mass, while asking God what He wants us to remember from His Word. We should also read the Bible everyday at home.

THE GOSPEL

The first four books of the New Testament are the four Gospels: Matthew, Mark, Luke, and John.We stand during the reading from the Gospel because it teaches us the words and deeds of our Lord Jesus Christ. As the Gospel begins, we make the Sign of the Cross on our forehead, lips, and heart. While doing so, we should silently ask the Lord to send His Holy Spirit to fill us with understanding and grace: "Lord, fill my mind, be on my lips and in my heart as I listen to Your Word, both now and forever."After the readings, there is a Homily, explaining God's Word, and General Intercessions for our needs.

PRESENTATION OF THE GIFTS

Jesus invites us to celebrate Mass with Him everyday. He waits for us, longing to give us the greatest gift of all – Himself. The bread and wine that we offer as gifts to God will become God's Gift to us. Through the priest, the bread and wine will be changed into Christ's Body and Blood to be given to us in Holy Communion. Can we give Jesus anything less than ourselves, by loving Him and being filled with thankfulness, especially when we receive Him in the Eucharist?

WASHING OF HANDS

The priest prepares for the celebration of the Eucharist by offering the gifts of bread and wine. After he does this, the priest washes his hands to get ready to touch the bread that will become Jesus Himself. Jesus is holy and must be touched with clean hands. As the servers pour water over his hands, he says, "Wash me O Lord, from my iniquity, and cleanse me from my sins." As the priest purifies his hands he then asks the Lord to make him pure inside as well so that he can worthily celebrate the Mass.

INCENSE AT THE ALTAR

The priest incenses the altar and the gifts that will be offered. Incense is a sign of our prayer rising up to God. It is at the beginning of Mass, the gospel, the offertory, and the consecration. Incense is not used at all Masses. It is the priest's choice whether to use it. Incense is mainly used at solemn high Masses and funerals.

PREPARATION OF THE ALTAR AND THE GIFTS

Food and drink are necessary to stay alive. Christ chose bread and wine to become His Body and Blood, to be Food and Drink for us. He is our life! As the bread and wine are placed on the altar, we are preparing for the most important part of Mass. We should be paying attention and asking Jesus to prepare us to receive Him in love.

HOLY, HOLY, HOLY

In the Eucharistic Prayer, the priest acts in the Person of Christ. In a prayer of praise and thanksgiving, he repeats the words and actions that Jesus said and did at the Last Supper. The Consecration makes miraculously present on the altar Christ in His Body and Blood, which He gave up for us at the Last Supper and at His Sacrifice on Calvary. Jesus becomes really present on the altar. We begin the Eucharistic Prayer joining the Saints and Angels singing to God's glory: "Holy, holy, holy..."

THE MYSTERY OF FAITH

In the Gospel of John, Jesus says, "Whoever eats my flesh and drinks my blood has eternal life, and I will raise him on the last day" (6:54). Jesus gives us His Body and Blood in the Eucharist. When we worthily receive Jesus in faith, He blesses us with His life and grace. As we receive Jesus, He makes us holy and helps us to be good. We call the Eucharist the great Mystery of Faith. As part of the Consecration Prayer, we join in one voice, saying the Memorial Acclamation: "Christ has died; Christ is risen; Christ will come again" to proclaim our belief in this Great Mystery.

DO THIS IN MEMORY OF ME

At the Last Supper, Jesus held the bread in His Hands and said:
"TAKE THIS, ALL OF YOU,AND EAT OF IT: THIS IS MY BODY..." Later, He took the wine and said: "TAKE THIS, ALL OF YOU, AND DRINK FROM IT: THIS IS THE CHALICE OF MY BLOOD..." Still later He said to His Apostles: "DO THIS IN MEMORY OF ME." These words and actions of Christ are repeated at every Mass, just as He commanded.
During the elevation of His Body and Blood, we adore and praise God, saying to ourselves: "My Lord and My God" or "Praise to You Lord Jesus Christ."

BELL RINGING WHEN THE PRIEST RAISES THE CHALICE

The chalice is being raised after it has been consecrated (blessed). The priest just said the words, "Do this in memory of me." With the conclusion of these words the cup of wine has been changed into the Blood of Jesus Christ. It is the same thing Jesus did at the Last Supper. At this time, the altar server rings the bells signifying that something very special has taken place. It is good to look at the chalice and say a prayer to Our Lord and thank him for all that he did for us and continues to do.

CONSECRATION OF THE HOST

The priest takes a piece of bread and says the words, "This is my Body, which will be given up for you." With these words ordinary bread becomes the true Body of Jesus Christ. It is a most sacred time of the Mass. He says same the words that Jesus said at the Last Supper. After the bread becomes Jesus, he raises the consecrated host and shows it to the people as in the same way as the chalice. When we gaze upon Jesus in the Eucharist, we say a prayer and thank the Lord for the gift of himself that makes us holy.

THE GREAT AMEN

As the priest says the words of the Consecration Prayer, the bread and wine become the Body, Blood, Soul and Divinity of Jesus Christ. The priest holds the Sacred Host and chalice up at the end of the Eucharistic Prayer to offer Jesus to the Father in Heaven on our behalf. This also reminds us of Christ's Death on the Cross and His Resurrection from the dead. At Calvary, Jesus won the victory over sin and death. We celebrate His victory at Mass as we pray with thankful hearts the Great 'Amen' because we believe in the salvation He won for us on the Cross, and we believe in the Real Presence of Jesus our Lord and Savior in the Eucharist.

THE LORD'S PRAYER

We have an Eternal Father in Heaven Who loves each one of us as if we were His only child. He created all things and gave us our earthly parents to love us and take care of us in His Name.When God the Son was asked how to pray, He taught His followers the Our Father prayer. Jesus called the Eternal Father "Abba" (Mark 14:36), which means "daddy." God wants us to love Him as a daddy too. He wants to take care of us and to give us all of His love and grace in the Eucharist. So, trusting in Him, we pray: 'Our Father...'

SIGN OF PEACE

Jesus is the Prince of Peace. We will only achieve true and lasting peace through Him. He shows us how to find peace, by telling us to love our neighbor unconditionally. Jesus teaches that when you "bring your gift to the altar, and there recall that your brother has anything against

you, leave your gift there at the altar, go first and be reconciled with your brother, and then come and offer your gift" (Matthew 5:23-24). When we turn to those around us at Mass to give and receive a sign of peace with them, we are also reconciling with the whole Church of Christ.

BREAKING OF THE BREAD

The priest breaks the Sacred Host at Mass like Jesus did at the Last Supper to remind us that His Body was broken for us on the Cross. The priest then puts a piece of the Host into the consecrated wine to show that Christ's resurrected Body and His Blood are reunited and now inseparable for all eternity. He died to destroy our death and He rose to restore our life. We are united to the resurrected Christ in Holy Communion. By receiving Jesus in the Eucharist, we share in His life now on earth and can anticipate when we will share in eternal life with Him forever in Heaven.

THE BLOOD OF CHRIST

Jesus says in John 15:13, "No one has greater love than this, to lay down one's life for one's friends." He gave up His life for us on the Cross, dying for our sins. He promises that all who believe in Him will have eternal life. We believe in You Jesus! Just as Jesus did at the Last Supper, the priest always receives the Sacred Host and the consecrated wine at Mass. For practical reasons the people sometimes receive Christ under the sign of bread alone. At other times, the people may receive both signs of bread and wine. In either case, they receive Christ fully in his Body, Blood, Soul and Divinity in each sign.

PRIEST RECEIVING COMMUNION

After the Lamb of God, the priest says a private prayer and reverently consumes the Eucharist. The priest must consume the Eucharist in order for the Mass to be valid. He is acting in the person of Jesus and doing what the Lord did the night he offered his last meal with his disciples. After he receives the consecrated host, the priest then drinks the Precious Blood from the chalice. This is also essential to the Mass. After that he distributes communion to the faithful in the congregation.

HOLY COMMUNION

The Mass is a sacred, holy meal. We do not eat to fill our physical bodies with ordinary food. We receive the Real Presence of Christ in this heavenly food to be more like Him and to share fellowship with the Father, Son, and Holy Spirit. When we receive the Eucharist, we are changed into what

we eat – Jesus Christ. "It is Christ in you, the hope for glory" (Colossians 1:27). He takes up His life within us and reigns in our hearts. After Communion, we should spend some time with Jesus in prayer, possibly saying some prayers found in this book.

AFTER RECEIVING HOLY COMMUNION

Holy Communion is God's great gift to his people. We thank him for his grace. Through Holy Communion, we receive God's very life, Grace, that changes us and makes us more into His image. After the faithful have received Holy Communion, they offer a prayer of thanksgiving to God the Father. God has blessed us in so many ways and we thank Him for His gifts.

CONCLUSION OF THE MASS

At the conclusion of Mass, just before they recess down the aisle, the priest and servers reverently bow to the altar and the Blessed Sacrament. As they walk down the aisle, the choir may sing a song. The recessional is a part of the Mass where the priest, in the place of Jesus Christ, slowly exits led by servers. They depart in the same order that they entered.

GO IN PEACE

Eucharist means "to give thanks". At Mass, we give thanks and praise to God for His many gifts and blessings. As the Mass ends, the priest asks the Father, Son and Holy Spirit to shower us with every grace and blessing from God. How very good it is to respond: "Thanks be to God." "God be with you" is a way of saying good-bye. As the Mass ends, we are sent forth, until we meet again at the next Mass and finally at last together in Heaven. Let us go in peace to love and serve the Lord and one another, taking all the graces and blessings we received to others. We should try to spend some time in thanksgiving with Jesus after Mass before we leave.

GREETING AFTER MASS

Many priests greet the congregation after Mass. Saying hello and goodbye to one another are basic Christian greetings. Christ was always ready to accept and greet all people. We are to do the same. After Mass is a good time to be happy and show our good will toward others. Many people come to the priest to wish him a good day or to thank him for the Mass. The priest also wishes people a good day and listens to what they have to say.

LIGHTING CANDLES BEFORE OR AFTER MASS

Lighting candles in a church is a devotion that expresses our love for the saints and for God. We are asking for the saints to intercede for us to the Lord. Their prayers are powerful and can help us lead good moral lives. Many people light candles to pray for themselves or loved ones.

THE SACRAMENT OF PENANCE

God will forgive all our sins if we are truly sorry, and go to Confession for the serious ones. The Sacrament of Reconciliation restores our friendship with Jesus and brings joy and peace to our heart.